CONSTRUCTION INDUSTRY TRAINING BOARD

STUDY NOTES

ELECTRICAL INSTALLATION

3

Basic Electrical Circuits

Revised and Reprinted 1989

EE 119/3

Published by

Construction Industry Training Board
Bircham Newton, Kings Lynn,
Norfolk PE31 6RH

Revised Edition 1989

ISBN 0 902029 52 5

CONTENTS

SAFETY NOTE

Electrical Isolation

Before beginning work on any electrical circuit you should make sure that it is completely isolated from the supply. Electrically powered machines are usually fitted with an isolator for disconnecting the supply under 'no load' conditions. Otherwise it may be necessary to isolate the supply by removing fuses, locking off MCBs or by physical disconnection of live conductors.

Fuses, switch fuses and isolators should be clearly marked to indicate the circuit they protect.

Any isolating device when operated, should be capable of being locked in the open position, or carry a label stating otherwise. If the isolator consists of fuses these should be removed to a safe place where they cannot be replaced without the knowledge of the responsible person concerned. For example, they can be kept in the pocket if the job is of short duration, or in a locked cupboard provided for the purpose in the charge of the works or site supervisor.

Fuses should **never** be removed or replaced without **first** switching off the supply.

Electrician's Responsibility

When working on a particular circuit or machine you must be certain that the supply cannot be switched on without your knowledge. You must satisfy yourself that the circuit is open and labelled so.

Always check that the circuit is dead, using approved test lamps, to prevent danger from electric shock (see diagrams overleaf).

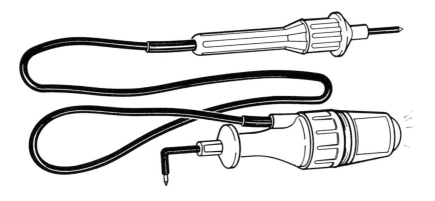

An Approved Test Lamp

1

ISOLATING A COMPLETE INSTALLATION

Flowchart 1

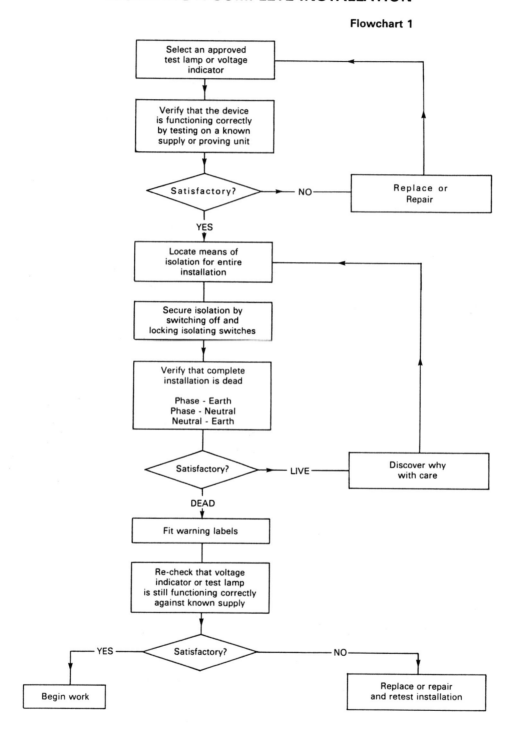

ISOLATING AN INDIVIDUAL CIRCUIT OR ITEM OF FIXED EQUIPMENT

Flowchart 2

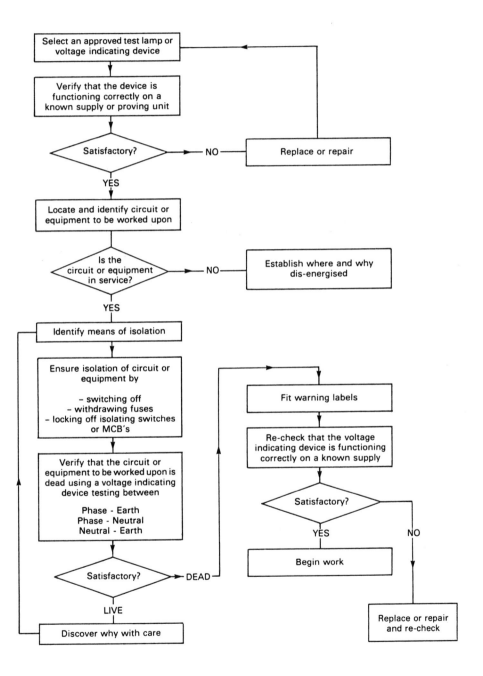

3

Test Lamps

Neon testers should not be relied on since the neon will not indicate supplies at low potentials. There is also the risk of receiving an electric shock if the resistor breaks down.

Flex and lampholder (homemade) test lamps should never be used. These are extremely dangerous, since mains potential is present in all the components; flex and bulb are vulnerable to damage, and there is usually no fuse in circuit.

Approved test lamps have a robust plastic body containing the circuitry. Resistors are fitted in the circuit to limit the potential and fuses are fitted to give complete protection in case of any fault occurring.

DRAWINGS AND CIRCUIT DIAGRAMS

A technical drawing or diagram is simply a means of conveying information more easily or clearly than can be expressed in words. In the electrical industry, drawings and diagrams are used in different forms. Most frequently used are

Block diagrams
Circuit diagrams
Wiring diagrams
Layout diagrams
As fitted drawings
Detail drawing

Block diagrams

The various items are represented by a square or rectangle clearly labelled to indicate its purpose.

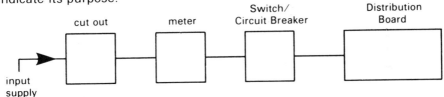

Circuit diagrams

A circuit diagram makes use of special symbols to represent pieces of equipment or apparatus and to show clearly how a circuit works. It may not indicate the most convenient way of wiring the circuit but it will show the electrical relationship between the various circuit elements.

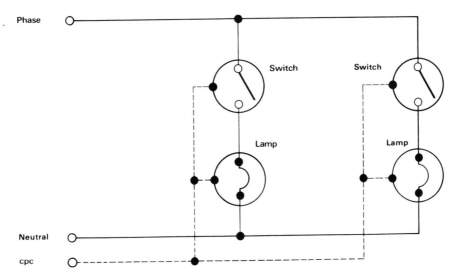

Wiring diagrams

Wiring diagrams give sufficient information for the connection of a circuit. In some respects wiring diagrams may be more detailed than circuit diagrams but they do not necessarily give any indication of how the equipment concerned operates.

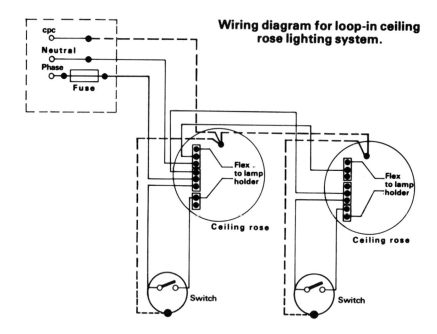

Wiring diagram for loop-in ceiling rose lighting system.

Layout drawings

These are scale drawings prepared under the supervision of an electrical consultant or engineer responsible for a particular installation and are based upon architects' drawings of the building in which the installation is to be effected. These drawings show the required position of all the equipment, metering and control gear to be installed.

Detail drawings

Sometimes additional drawings will be provided to clarify points of detail. These are 'detail drawings'. A typical example would be a drawing of the construction and fixing of a main switchgear panel.

'As fitted' drawings

Once an installation is completed a set of drawings should be produced indicating any modification of the original layout. These should be retained as a record and are frequently referred to as 'fitted' drawings.

Electrical systems depend on there being a complete circuit from the source of supply to the apparatus and back again. Every circuit must start from one pole of the supply and return to another pole (eg. from phase to neutral). In an installation supplied in the ordinary way from the Electricity Board's mains, the source of supply is the supply terminals on the consumer's fuseboard.

Circuits are interrupted so that the power may be controlled by

- switches
- fuses
- contactors and relays
- time switches
- thermostats

PLANNING FINAL CIRCUITS

A final circuit is that part of an installation between an item of current consuming equipment and the fuse or circuit breaker at the consumer unit. It is in effect the point where load is applied to a system. The design of final circuits is very important and every precaution must be taken to ensure they do not constitute a hazard to the user.

Circuit Design

There are many factors which have to be considered in the design of final circuits. Some are purely electrical – others are concerned with location.

Consider first the electrical factors. The rating of the cables, switches and protective devices must be decided together with the form and size of circuit protective conductor and the type of protection to be used.

Physical factors such as the distance from the supply to the load which could introduce a voltage drop must also be considered. Specialised hazards such as the danger of fire and explosion will affect the type of switchgear, cable and other equipment required (as in spray paint shops and flour mills). The temperature, and the proximity to other cables and thermal insulation are all factors which would require larger cables to be selected for the given load.

LIGHTING CIRCUITS

ONE-WAY LIGHTING CIRCUIT

The simplest circuit consists of a pair of wires from the mains terminals supplying a lamp. In this circuit there must be a switch which (if single pole) must be situated in the phase conductor.

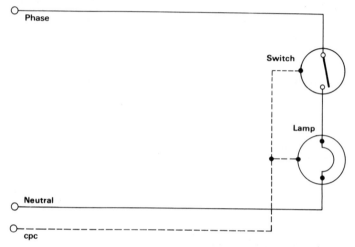

TWO-WAY LIGHTING CIRCUIT

A two-way lighting circuit is often used on staircases so that one can switch off the downstairs light from upstairs, or vice versa.

In this circuit, the switches can have two positions, either of which can light the lamp. Suppose switch A is in the upper position, and switch B is in the lower position, as illustrated, there is no circuit, so the lamp is out.

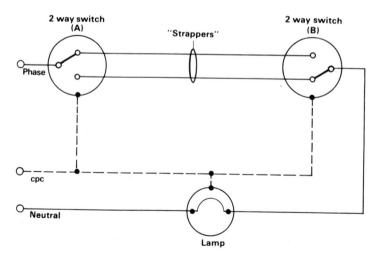

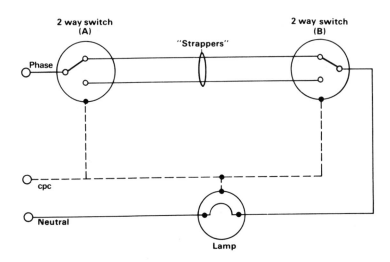

If switch B is operated a circuit is established, and the lamp lights. Now if switch A is operated the lamp goes out. The two wires between switches A and B are called 'strappers'.

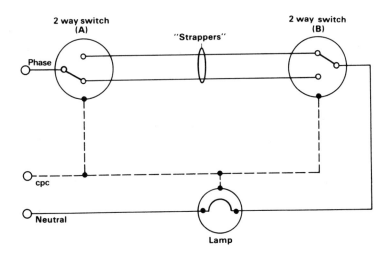

Intermediate lighting circuit

When it is required to control the lighting in a long corridor or staircase with several landings, it is desirable to arrange for lights to be switched on and off at several points. In this case intermediate switches should be used.

If the two wires between the switches A and B as illustrated (the strapping wires) were reversed, a circuit would be established and the lamp would light.

The intermediate switch carries out this reversal of the strapping wires. Any number of intermediate switches may be installed.

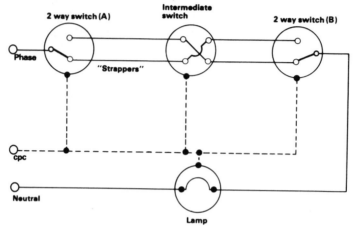

Conversion of one-way lighting circuit to two-way switching

It is sometimes necessary to modify lighting circuits controlled from one point.

Consider an existing one-way circuit such as that previously illustrated. To convert it to a conventional two-way circuit, either the switch feed, or the switch wire must be removed, and replaced by a conductor going to another switch position. This alteration must be made either at the lamp or at the consumer unit.

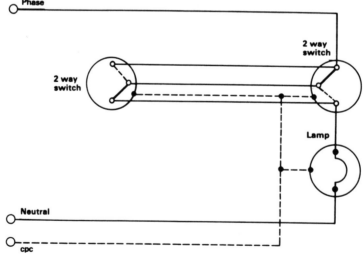

This can be avoided by using the circuit illustrated. Two-way switch control can be achieved by replacing the original switch by a two-way switch connected as illustrated and by running three new wires to a two-way switch at the new control position.

Joint box method

For two lights independently controlled the circuit would be as illustrated.

1. Joint box showing phase, neutral and circuit protective conductor (c.p.c).

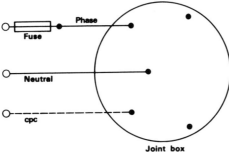

Joint box

2. Switch connected showing switch feed and switch wire and circuit protective conductor (c.p.c.)

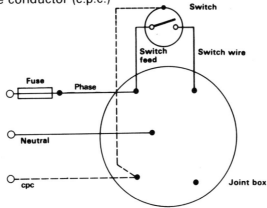

3. Light connected showing switch wire and neutral and circuit protective conductor (c.p.c.)

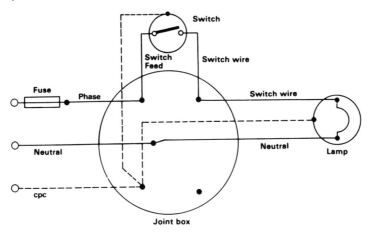

4. Additional light connected showing all switch feeds, switch wires, neutrals and circuit protective conductors (c.p.c.)

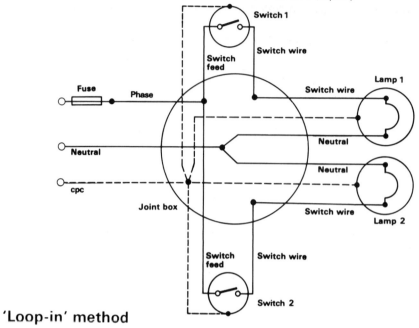

'Loop-in' method

The most common system of wiring final sub-circuits is the loop-in system where all connections are made at the electrical accessories.

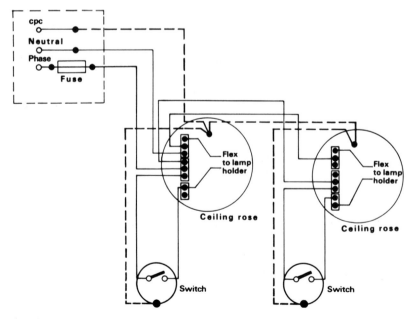

For simplicity all wiring diagrams have to show the basic circuit wiring necessary for the circuit. The loop-in system of wiring at the ceiling rose would be as illustrated.

Number of points

The number of points which may be supplied by a final circuit not exceeding 15A rating is limited by their aggregate demand as determined by Table 4A of the IEE Regulations.

TABLE 4A

Current demand to be assumed for points of utilisation and current-using equipment

Point of utilisation or current-using equipment	Current demand to be assumed
Socket outlets other than 2A socket outlets	Rated current
2A socket outlets	at least 0.5A
Lighting outlet*	Current equivalent to the connected load, with a minimum of 100W per lampholder
Electric clock, electric shaver supply unit (complying with BS 3052), shaver socket outlet (complying with BS 4573), bell transformer, and current-using equipment of a rating not greater than 5VA	May be neglected
Household cooking appliance socket outlet is incorporated in the control unit	The first 10A of the rated current plus 30% of the remainder of the rated current plus 5A if a
All other stationary equipment	British Standard rated current, or normal current

Note: Final circuits for discharge lighting are arranged as so to be capable of carrying the total steady current, viz. that of the lamp(s) and any associated gear and also their harmonic currents. Where more exact information is not available, the demand in volt-amperes is taken as the rated lamp watts multiplied by not less than 1.8. This multiplier is based upon the assumption that the circuit is connected to a power factor of not less than 0.85 lagging, and takes into account control gear losses and harmonic currents

Generally for domestic lighting circuits the number of lights is limited to a maximum of ten.

Cable size and overcurrent protection

Cable size for wiring domestic lighting circuits is generally 1 mm² twin with c.p.c. PVC cable; the protective device being limited to a maximum of 6A.

Flexible cord pendant

A ceiling rose must not be used for more than one flexible cord unless it is specially designed to take multiple pendants.

LIGHTING CIRCUITS IN INDUSTRIAL INSTALLATIONS

It is often necessary and desirable to rate final lighting circuits up to 16 amperes. The reason for this is that lamps of 500-1,000 watts may be used; several wired on one circuit with 2.5mm^2 cable. In such installations Edison screw lampholders should be used.

When planning an industrial lighting installation it is not always possible to use the 'diversity factor' as permitted by IEE Regulations because the lights on these circuits are often all switched on at one time over long periods. This illustrates the need for knowledge at the planning stage of the use to which the installation will be put.

POWER CIRCUITS

13A SOCKET OUTLET CIRCUITS

Prior to the early 1950's the multiplicity of types and sizes of socket outlet and plugs in domestic premises had always been a source of annoyance to the householder with 5A and 15A socket outlets installed. In 1947, agreement was reached on a standard socket outlet with a fused plug (BS 1363).

The advantages of this socket outlet are that an appliance can be used in any room; the cost of the wiring is kept as low as possible, whilst affording proper protection for each appliance.

The 13 amp socket outlet system is based on the principle of 'diversity of use'.

Standard circuit arrangements

Types of final circuit using BS 1363 socket outlets and fused connection units are :-

- Ring circuits

- Radial circuits

THE RING CIRCUIT

In this system the phase, neutral and circuit protective conductors are connected to their respective terminals at the consumer unit and loop into each socket in turn; and then return to their consumer unit terminals, forming a 'ring'. Each 13A socket has two connections back to the mains – each capable of carrying 13A at least.

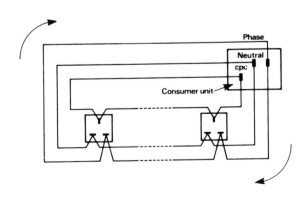

**Basic ring final
circuit
diagram**

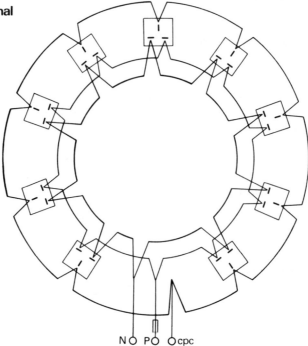

N◯ P◯ ◯cpc

Each ring final circuit conductor must be looped into every socket outlet or joint box which forms the ring and must be electrically continuous throughout its length.

The requirements for ring circuits are:-

- An unlimited number of socket outlets may be provided. (Each socket outlet of a twin or multiple socket to be regarded as one socket outlet).

- The floor area served by a single 30A ring circuit must not exceed 100m^2 in domestic installations.

- Consideration must be given to the loading of the circuit especially kitchens which may require a separate circuit.

- When more than one ring circuit is installed in the same premises, the socket outlets installed should be reasonably shared amongst the ring circuits so that the assessed load is balanced.

Cable sizes and overcurrent protection are given in Table 5A Appendix 5 of the IEE Regulations. (see overleaf)

Final circuits using BS 1363 socket outlets

Type of circuit	Overcurrent protective device		Minimum conductor size*			Maximum floor area served
			Copper conductor rubber- or p.v.c. insulated cables	Copperclad aluminium conductor p.v.c. insulated cables	Copper conductor mineral- insulated cables	
1	Rating 2	Type 3	4	5	6	7
A1 Ring	A 30 or 32	Any	mm² 2.5	mm² 4	mm² 1.5	mm² 100
A2 Radial	30 or 32	Cartridge fuse or circuit breaker	4	6	2.5	50
A3 Radial	20	Any	2.5	4	1.5	20

The tabulated values of conductor size may be reduced for fused spurs.

Spurs to a ring circuit

Spurs may be installed by a ring circuit. These may be fused – but it is more common to install non-fused spurs connected directly to the ring circuit conductors. Fused spurs are connected via a protective device such as a fuse.

The total number of fused spurs is unlimited, but the number of non-fused spurs must not exceed the total number of socket outlets and any stationary equipment connected directly to the circuit. A non-fused spur may supply only one single or one twin socket outlet or one item of permanently connected equipment.

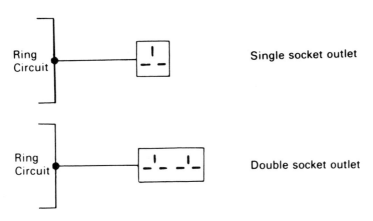

Ring Circuit — Single socket outlet

Ring Circuit — Double socket outlet

TYPICAL RING CIRCUIT

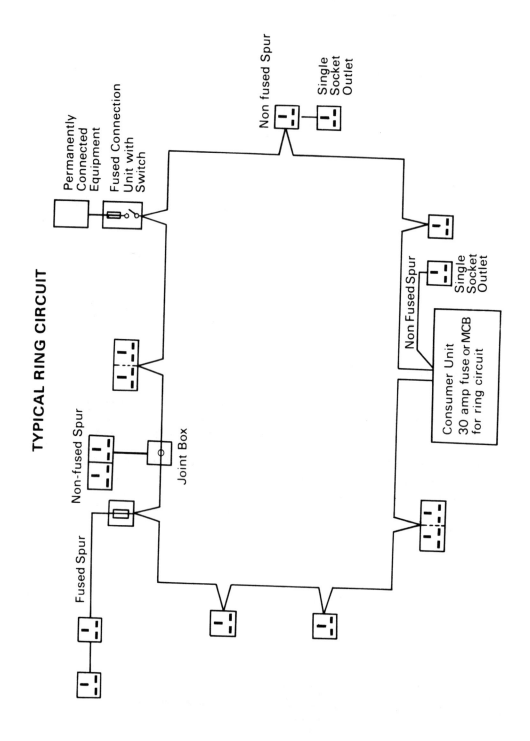

Permanently Connected Equipment

Fused Connection Unit with Switch

Non fused Spur

Single Socket Outlet

Non-fused Spur

Joint Box

Fused Spur

Non Fused Spur

Single Socket Outlet

Consumer Unit 30 amp fuse or MCB for ring circuit

18

Permanently connected equipment

Permanently connected equipment should be locally protected by a fuse (not exceeding 13A) and be controlled by a switch complying with the Regulations, or be protected by a circuit breaker not exceeding 16A rating.

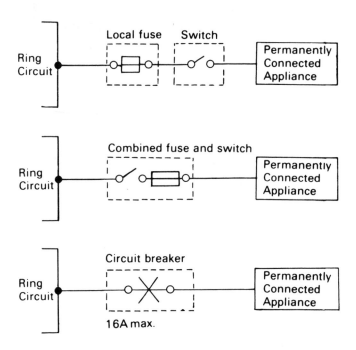

Note
The cable sizes of spurs should not be less than that of the ring circuit

Fused spurs

A fused spur is connected to a circuit through a fused connection unit. The fuse incorporated should be related to the current carrying capacity of the cable used for the spur, but should not exceed 13A.

When socket outlets are wired from a fused spur the minimum size of conductor is:

$1.5mm^2$ for rubber or PVC insulated cables with copper conductors.

$2.5mm^2$ for rubber or PVC insulated cables with copper clad aluminium conductors.

$1.0mm^2$ for mineral insulated cables with copper conductors.

FUSED SPURS

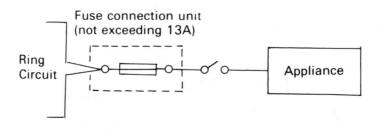

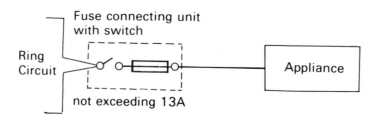

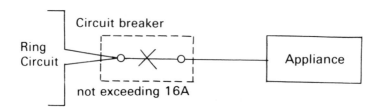

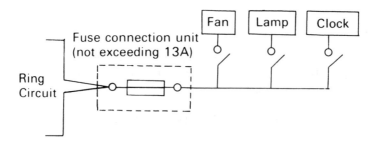

Note

**Cable size for spur is dependent on the
magnitude of the connected load.**

Method of Connecting Spurs to Circuit

(a) at the terminal of accessories on the ring

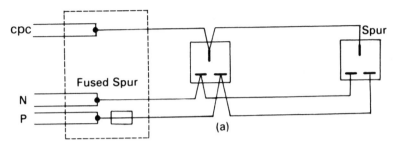

(a)

(b) at a joint box

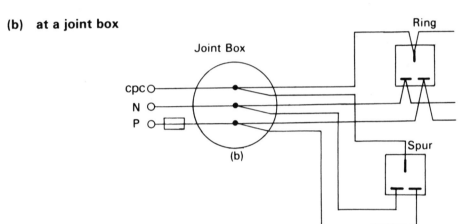

(b)

(c) at the origin of the circuit in the distribution board

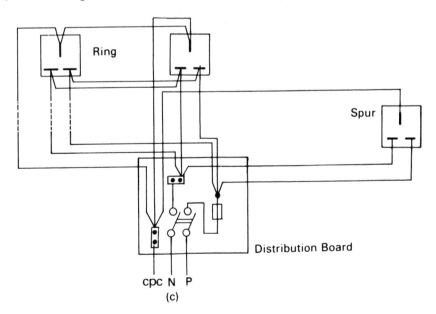

(c)

RADIAL CIRCUITS *(Appendix 5)*

Radial circuits also make use of 13A sockets (BS 1363) but the circuit is not wired in the form of a ring.

An unlimited number of socket outlets may be supplied, but the floor area which may be served by the socket outlets is limited to either 20m² or 50m² depending on the size and type of cable used and size of overcurrent protection afforded.

FIXED EQUIPMENT

Where immersion heaters are installed in storage tanks with a capacity in excess of 15 litres, or a comprehensive space heating installation is to be installed, for example, electric fires or storage radiators, separate circuits should be provided for each heater.

Immersion heaters

Cable protection — damage by heat

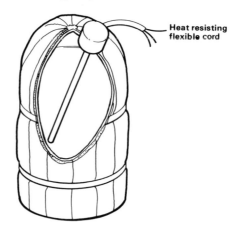

Heat resisting
flexible cord

With a permanently installed hot water system, the heater is placed in a cylinder or tank. Hot water is drawn off at sink, basin or bath.

Storage cylinders and tank systems all use electric heating elements immersed in the water, allowing for an efficient transfer of heat. They are generally thermostatically controlled.

The immersion heater must be wired on a separate radial final circuit.

The flexible cord used to make the final connection between the circuit and the immersion heater may reach quite a high temperature, especially if lagging materials should cover it. In these circumstances a heat-resistant flexible cord should be used.

Storage water heaters

There are various models available and these vary considerably in size, shape and capacity.

The cylinder or tank does not hold water at pressure. With this type of heater no plumbing will be required except for the main cold water feed. Other models require no plumbing. These are fitted with a flexible hose which fits onto an existing tap.

Electrical connections to storage water heaters must be by way of a heat resisting flexible cord.

A typical final circuit diagram is illustrated below.

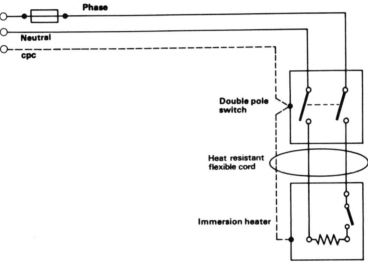

Dual immersion heater circuits

It is possible to fit two immersion heaters with thermostats into the same cylinder to achieve either a full, or a part cylinder of hot water.

A typical dual immersion heater circuit diagram is illustrated.

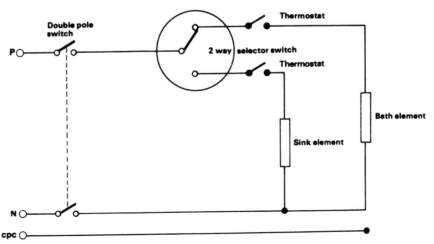

Cooker circuits

A circuit supplying a cooking appliance must include a control switch or cooker control unit (which may include a socket outlet). The rating of the circuit should be determined by an assessment of the current demand in accordance with Table 4A Appendix 4 of the Regulations.

For a circuit where the rating exceeds 15A but does not exceed 50A, two or more cooking appliances which are installed in the same room may be supplied, e.g. split level cookers. However, where either the cooker or the hob is installed in excess of 2 metres from the control unit a separate means of isolation must be installed for each appliance.

The method of wiring for a standard cooker is as illustrated. A cable is run direct from the consumer unit to a cooker control unit or double switch.

Wiring diagram for an electric cooker.

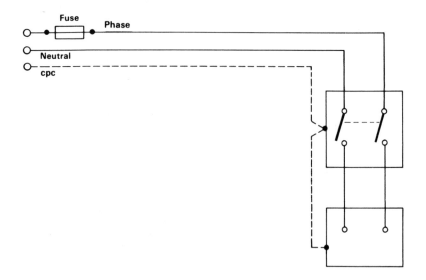

When wiring a 'split level' cooker a cable is run from the consumer unit to a double pole isolating switch. If the hob and oven units are within 2 metres of the isolating switch then the two units are wired from this switch. The wiring diagram is as illustrated.

24

Control of stationary cooking appliances in domestic premises.

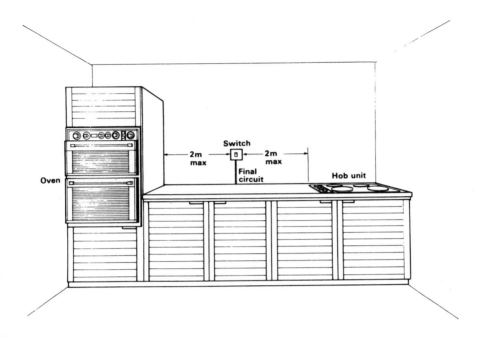

Wiring diagram for split level electric cooker.

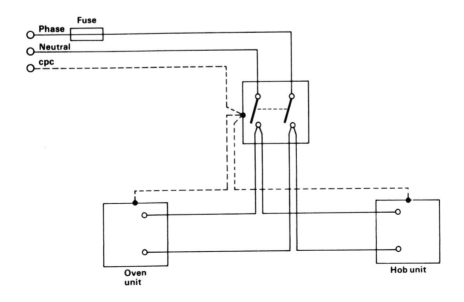

CONTROL AND PROTECTION EQUIPMENT

THE ELECTRICITY BOARD'S EQUIPMENT

The Central Electricity Generating Board is responsible for the generation and transmission of electricity, but the responsibility for its distribution to consumers is shared by twelve Electricity Boards. Consumers' installations are connected to the Electricity Board's cables by means of service cables, and these are terminated within the premises at a mutually agreed position.

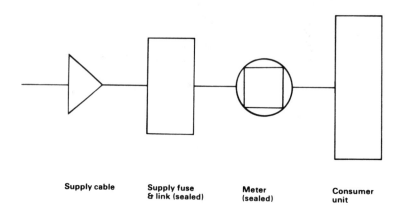

| Supply cable | Supply fuse & link (sealed) | Meter (sealed) | Consumer unit |

At the service termination point the service cable is connected to a service fuse. This fuse, which normally has a rating of 80 or 100A, will ensure that no damage occurs to the Board's cables or equipment in the event of a fault occurring on the consumer's premises. Connections are installed by the Board from the service fuse to the meter, which is supplied to record the amount of electricity consumed. The service fuse and meter are sealed by the Electricity Board and consumers or contractors are not permitted to interfere with this part of the installation.

THE CONSUMER'S INSTALLATION

Connections are made from the meter to one or more main switches, and it is at this point that the consumer's responsibility for the installation begins. Main switches must be provided so that the installation may be isolated from the supply when alterations or extensions are made.

Consumer's control unit

For domestic installations where the load does not exceed 100 amperes the use of a consumer's control unit is recommended.

Consumer's control units are made to BS 1454 and usually consist of a 60-80 ampere double pole switch and a number of fused outlets. An arrangement with eight fuse ways may be used to supply the following circuits:

Two 30 ampere ways for two ring final circuits

One 30 ampere way for the cooker circuit

One 20 ampere way for the garage power circuit

One 15 ampere way for the immersion heater circuit

Two 5 ampere ways for the lighting circuits

One spare way

There will be many variations; no two types of installation will have exactly the same needs.

Each consumer unit will have its own double pole main switch, fuseholders or miniature circuit breakers for each way, neutral terminals and earth terminals.

Metal clad or insulated enclosures may be installed to accommodate the consumer unit. If a conduit system is being installed a metal clad fuseboard will be used. For sheathed cable systems an insulated enclosure is more usual. A consumer unit constructed without a back must be mounted on a non-flammable surface or fitted with a non-flammable backing plate.

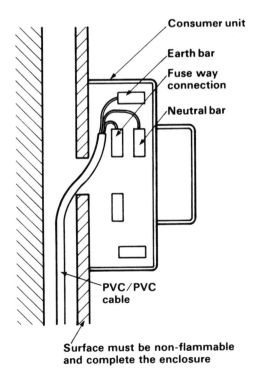

Consumer unit

Earth bar

Fuse way connection

Neutral bar

PVC/PVC cable

Surface must be non-flammable and complete the enclosure

Main switchgear

Every installation must be controlled by one or more main switches. The main switchgear may consist of a switchfuse or a separate switch and fuses and must be readily accessible to the consumer and as near as possible to the Electricity Board's service intake. The Factories Act states that *'an efficient means, suitably located shall be provided for cutting off all pressure from every part of a system, as may be necessary to prevent danger.'*

The general requirements for isolation and switching in Chapter 46 of the IEE Regulations states that *'every installation and all items of equipment should be provided with effective means to cut off all sources of voltage.'*

CONTROL OF SEPARATE INSTALLATIONS

Control apparatus must be provided for every section of a consumer's installation. An off-peak supply, such as that supplying electric storage heaters, is considered to require separate metering and control apparatus in addition to the metering and control apparatus for lighting and other loads.

Where a consumer's installation is split up into separately controlled parts, each part must be treated as a separate installation. This applies irrespective of whether the parts are within the same building or in separate detached buildings.

Every installation must be controlled by one or more main switches. The main switchgear may consist of a switchfuse or separate switch and fuses.

Main switchgear. Permissible combinations

Layout for combined lighting & heating tariff metering

Layout when lighting & heating are separately metered

Control of detached buildings

Installations in detached buildings must each be provided with a specified means of isolation for example as illustrated.

Separate isolation in detached buildings

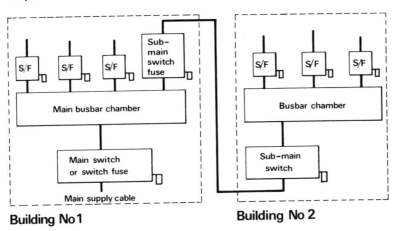

Building No 1

Building No 2

IDENTIFICATION NOTICES

Switchgear control gear and protective devices

Switchgear and control gear in an installation should be labelled to indicate its use. All protective devices in an installation should be arranged and identified so that their respective circuit may be easily recognised.

Diagrams and charts must be provided for every electrical installation indicating:

(a) the type of circuits

(b) the number of points installed

(c) the number and size of conductors

(d) the type of wiring system

(e) the location and types of protective devices and isolation and switching devices

(f) details of the characteristics of the protective devices for automatic disconnection, the earthing arrangements for the installation and the impedences of the circuit concerned (413-3).

Note: For simple installations the foregoing information may be given in a schedule, if symbols are used they should conform to BS 3939.

The purpose of providing diagrams, charts and tables for an installation is so that it can be inspected and tested in accordance with Part 6 of the Regulations and to provide any new owner of the premises (should the property change hands) with the fullest possible information concerning the electrical installation.

It is essential that diagrams, charts and tables are kept up to date.

A typical chart for a small installation is illustrated below.

TABLE 16

Schedule of installation at . . . .
as prescribed in the IEE 'Regulations for electrical installations'

Type of circuit	Points served	Phase Conductor mm^2	Protective Conductor mm^2	Protective and switching devices
Ring final circuit	General purpose 13A socket outlets	2.5	1.5	30A miniature circuit breaker; type 2 local switches and plugs and socket outlets
No.1 lighting circuit	Downstairs fixed lighting	1.5	1.0	5A mcb; type 2 local switches
No.2 lighting circuit	Upstairs fixed lighting	1.5	1.0	5A mcb; type 2 local switches
Cooker circuit	Cooker	6.0	2.5	45A mcb; type 2 cooker control unit
Immersion heater circuit	Immersion heater	2.5	1.5	15A mcb; type 2 local switched connector box

Type of wiring: PVC insulated and sheathed, flat twin with cpc BS 6004

ACCESSORIES

Plugs and socket outlets

Plugs and socket outlets are designed so that it is not possible to engage any pin of the plug into a live contact of a socket outlet whilst any other pin of the plug is exposed, and the plugs are not capable of being inserted into sockets of systems other than their own. Where plugs containing a fuse are required, they must be non-reversible and arranged so that the fuse cannot be connected in the neutral conductor.

Plugs and socket outlets other than those specified in Table 55A of the IEE Regulations (below) may be used on single phase circuits operating at voltages not exceeding 250V for the connection of:

Electric clocks	–	use clock connector unit incorporating a BS 646 or 1362 fuse not exceeding 3 amperes.
Electric shavers	–	Use BS 4573
Electric shavers	–	use BS 3052 shaver unit for use in bathrooms.

The following types of plugs and socket outlets are recognised as being suitable for electrical installations by the IEE Regulations for low voltage circuits.

TABLE 55A

Plugs and socket outlets for low voltage circuits

Type of plug and socket outlet	Rating (amperes)	Applicable British Standard
Fused plug and shuttered socket outlets, 2 pole and earth, for a.c.	13	BS 1363 (fuse to BS 1362)
Plugs, fused or non-fused, and socket outlets, 2-pole and earth	2,5,15,30	BS 546 (fuse, if any, to BS 646)
Plugs, fused or non-fused, and socket outlets, protected type, 2-pole with earthing contact	5,15,30	BS 196
Plugs and socket outlets (theatre type)	15	BS 5550
Plugs and socket outlets (industrial type)	16,32,63,125	BS 4343

Electrical Accessories

13A BS 1363
Socket Outlet

Use with fuses
to BS 1362

Shaver Unit to
BS 4573
(Not for use in
bathrooms).

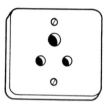

2, 5, 15 and 30A
Socket Outlets to
BS 546

Use with fuse to
BS 646 when
necessary.

Shaver Unit to
BS 3052
(For use in
bathrooms).

Incorporates
isolating
transformer

Clock Connector.
Use fuse not
exceeding 3A
to BS 646 or 1362

BS 4343 Plugs, Socket Outlets, Cable Couplers and Inlets

BS 196 Plug Socket and Couplers

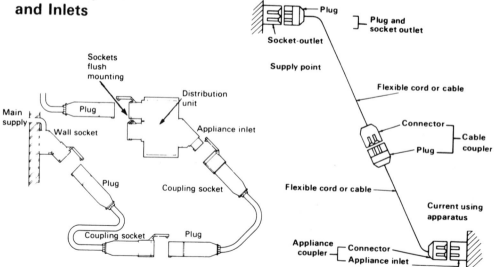

MAXIMUM DEMAND AND DIVERSITY

DIVERSITY

Consider a domestic installation. It is extremely unlikely that all appliances and equipment will be in full use at any time; for example, in normal circumstances a householder would be unlikely to switch on all the appliances – kettle, fires, water heaters, iron, toaster and cooker – at the same time, and it would be uneconomical to provide cables and switchgear of a capacity for the maximum possible load; the loads they will carry are likely to be less than the maximum. It is this factor which is referred to as 'Diversity'. By making allowances for Diversity the size and cost of conductors, protective devices and switchgear can be reduced.

To calculate the 'Diversity Factor' $\frac{\text{minimum actual load}}{\text{installed load}}$ for every type of electrical installation specialist knowledge and experience is required. Table 4B (Appendix 4) of the Regulations lists allowances for Diversity for final lighting, power, cooking, etc. circuits, for installations in household and small commercial premises such as shops, offices and guest houses. The use of Table 4B can result in the choice of switchgear and cables of a lesser current rating which might reduce the cost of an installation.

The common methods of obtaining the current demand of a circuit is to add together the current demand of all points of utilisation and equipment in a circuit. Typical current demand for points of utilisation and equipment are given in Table 4A (page 111) of the Regulations.

TABLE 4A
Current demand to be assumed for points of utilisation and current-using equipment

Point of utilisation or current-using equipment	Current demand to be assumed
Socket outlets other than 2A socket outlets	Rated current
2A socket outlets	at least 0.5A
Lighting outlet*	Current equivalent to the connected load, with a minimum of 100W per lampholder
Electric clock, electric shaver supply unit (complying with BS 3052), shaver socket outlet (complying with BS 4573), bell transformer, and current-using equipment of a rating not greater than 5VA	May be neglected
Household cooking appliance	The first 10A of the rated current plus 30% of the remainder of the rated current plus 5A if a socket outlet is incorporated in the control unit
All other stationary equipment	British Standard rated current, or normal current

Note: Final circuits for discharge lighting are arranged as so to be capable of carrying the total steady current, viz. that of the lamp(s) and any associated gear and also their harmonic currents. Where more exact information is not available, the demand in volt-amperes is taken as the rated lamp watts multiplied by not less than 1.8. This multiplier is based upon the assumption that the circuit is corrected to a power factor of not less than 0.85 lagging, and takes into account control gear losses and harmonic currents.

Example — Household Cooking Appliances

Application of Diversity Factor

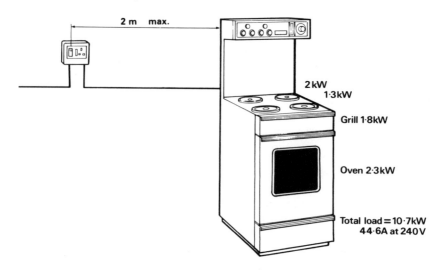

2 m max.

2 kW
1·3kW

Grill 1·8kW

Oven 2·3kW

Total load = 10·7kW
44·6A at 240V

If we consider an electric cooker with a maximum loading of 44.6 amperes, as illustrated, the assessed current demand would be as follows:-

The first 10 amperes of the total rated current of the cooker, *plus* 30% of the remainder of the total rated current of the cooker, *plus* 5 amperes if a socket outlet is incorporated in the control unit.

Total current rating		44.6A
First 10 amperes	=	10A
30% of remaining 34.6	=	10.38A
Socket outlet	=	5A
Assessed current demand		25.38A

Discharge lighting

Final circuits supplying discharge lighting should be capable of carrying the total steady current of the lamp and associated control gear. Where exact information is not available, provided the power factor of the circuit is not less than 0.85, the current demand of a discharge lamp can be calculated from the wattage of the lamp, multiplied by 1.8.

Therefore steady current of a discharge lamp =

$$\frac{\text{lamp power (watts)} \times 1.8}{\text{supply voltage}}$$

Example

A circuit supplies five, 240 volt single phase fluorescent fittings, each rated at 65 watts. The current demand would be:

$$I = \frac{5 \times 65 \times 1.8}{240} = 2.43A$$

Socket outlets

For standard power circuits using BS 1363 socket outlets detailed in Appendix 5 (page 113), no Diversity allowance should be made since this has already been taken into account.

Methods of applying diversity

For the design of an installation one of the following methods may be used.

Method 1

The assessed current demand of a circuit supplying a number of final circuits can be obtained by adding the current demands of all equipment supplied by each final circuit of the system and applying the allowances for Diversity given in Table 4B of the IEE Regulations. For a circuit with socket outlets wired in accordance with Appendix 5, the rated current of the protective devices is the current demand of the circuit.

Method 2

The alternative method of assessing the current demand of a circuit supplying a number of final circuits is to calculate the rating for each circuit, applying the allowances (Table 4B) and then add the current demand of the individual circuits together. A further allowance for Diversity may be applied, on the assumption that not all the circuits will be in use at the same time. The allowances given in Table 4B are not to be applied to the Diversity between circuits. The values used should be chosen by the designer of the installation. This method applies the principle of Diversity to the circuits as well as to connected equipment and appliances.

Method 3

The current demand of a circuit determined by a suitably qualified Electrical Engineer.

Example

Consider a small guest house with 10 bedrooms, 3 bathrooms, lounge, dining room, kitchen and utility room with the following loads connected to three phase 415/240V supply.

Lighting   3 circuits tungsten lighting.   Total 2,860 watts

Power   3 × 30A ring circuits to 13A socket outlets

Water heating   1 × 7 KW shower
2 × 3 KW immersion heater thermostatically controlled

Cooking appliances   1 × 3 KW cooker
1 × 10.7 KW cooker

Calculations and answer to example

		Current Demand (Amperes)	Table 4B (Diversity Factor)	Current Demand allowing for Diversity (Amperes)
Lighting		$\dfrac{2,860}{240}$ 11.92	75%	8.94
Power	(i)	30	100%	60
	(ii)	30	50%	
	(iii)	30	50%	
Water Heaters (inst)		$\dfrac{7,000}{240}$ 29.2	100%	29.2
Water Heaters (thermo)		$\dfrac{6,000}{240}$ 25	100%	25
Cooker	(i)	$\dfrac{10,700}{240}$ 44.58	100%	44.58
	(ii)	$\dfrac{3,000}{240}$ 12.5	80%	10

Total Current Demand (allowing Diversity)   177.72

Load – spread over 3 phases

$$= \frac{177.72}{3}$$

$$= 59.24A$$

$$= 60A \text{ per phase}$$

PROTECTION AGAINST SHOCK

ELECTRIC SHOCK

The severity of an electric shock depends on the level of electric current which passes through the body, and the duration of the contact.

At low levels of current the effect may be only an unpleasant tingling sensation, but this in itself may be enough to cause someone to lose their balance, and fall. Medium levels of shock can cause muscular tension, so that anything grasped can scarcely be released, making it difficult for the victim to disengage the contact. High levels of electric shock cause fibrillation of the heart, which is almost invariably lethal. Electric shock can also cause burning of the skin at the points of contact.

Electric shock may result from *direct contact;* that is contact with a live part, such as an exposed phase conductor.

Electric shock may also result from *indirect contact*; for example, by contact with the exposed metal framework or casing of an appliance, which should not normally be live, but which has become so because of a fault.

PROTECTION AGAINST DIRECT CONTACT

Protection against direct contact is provided by;

- the insulation of live parts.

- the provision of barriers or enclosures

- the placing of obstacles of one kind or another

- placing live parts out of reach.

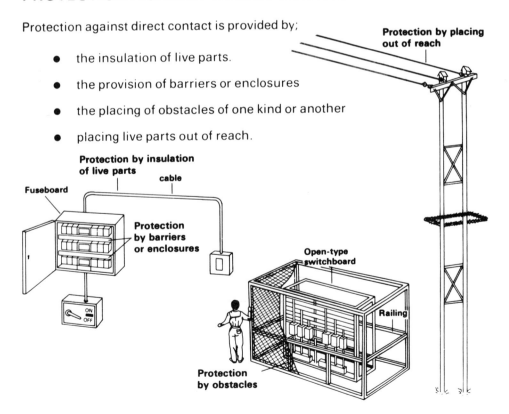

Protection by placing out of reach

Protection by insulation of live parts

cable

Fuseboard

Protection by barriers or enclosures

Open-type switchboard

ON OFF

Railing

Protection by obstacles

37

PROTECTION AGAINST INDIRECT CONTACT

Earthed Equipotential Bonding and Automatic Disconnection of Supply

When a phase to earth fault of negligible impedance occurs the potential of the metalwork is raised to a level which is likely to be dangerous.

The most common method of protection against indirect contact is that known as 'earthed equipotential bonding and automatic disconnection of the supply'. This relies on the fact that all metalwork of the system is connected to earth, and when an earth fault occurs the fault current is sufficient to operate a fuse or miniature circuit breaker.

The key requirement of this protective measure is the **RAPID** disconnection of the supply in the event of an earth fault. Regulation 413-4 of the IEE Wiring Regulations stipulates the maximum disconnection time:-

- for final circuit supplying socket outlets disconnection must occur within 0.4 seconds

- for final circuits supplying fixed equipment disconnection must occur within 5 seconds.

An exception to the 5 sec. time limit for final circuits supplying fixed equipment is stipulated for circuits supplying equipment in a room with a fixed bath or shower. For such circuits the disconnection time must be within 0.4 seconds.

When an earth fault occurs in an installation the degree of risk is greater with portable equipment held in the hand, such as a portable drill or sander. The flexible cables and cords which are used with portable equipment are subjected to wear and tear and there is no control over the lengths used.

For equipment to be used outside the Equipotential Zone

A further set of conditions has to be applied for equipment to be used outside the zone in which the circuit feeding that equipment originated (eg. electric lawn mowers and hedge trimmers supplied from socket outlets, electric pumps in certain industrial processes).The IEE Wiring Regulations specify that such equipment must be protected by a residual current device (r.c.d.) with an operating current not exceeding 30mA. In domestic premises the location of such an outlet would typically be in the garage.

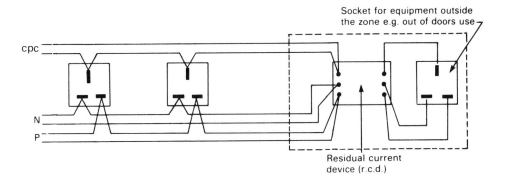

Socket for equipment outside the zone e.g. out of doors use

cpc

N

P

Residual current device (r.c.d.)

Earth Fault Loop Impedance

Where the disconnection times referred to earlier are to be achieved by means of overcurrent protective devices, it is necessary to relate the earth fault loop impedance to the type and rating of protective device used to protect the circuit. In order to determine the maximum earth fault loop impedance for a particular circuit it is necessary to refer to the time/current characteristic for that particular device.

Tables 41A1 *(Socket outlet circuits)* and 41A2 *(Fixed equipment)* give the limiting values of earth fault loop impedance appropriate to various types and ratings of protective device. The tabulated values of impedance are based on the assumption that the earth fault is of negligible impedance.

The earth fault loop impedance (Z_S) is made up of the impedance of the consumers phase and protective conductors (R_1 and R_2 respectively), and the impedance external to the installation (Z_E) the impedance of the supply. As the value of Z_E will be obtained from the Area Board for the initial assessment of the installation, the maximum impedance allowed for the phase and protective conductors can be determined from:

$$R_1 + R_2 = Z_S - Z_E$$

This limit on the value $R_1 + R_2$ must now effectively limit the length of run for the circuit cable.

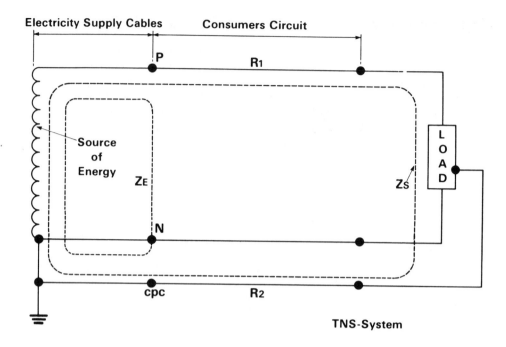

Note: *Appendix 17 of the IEE Wiring Regulations gives values for the resistance per metre of* $R_1 + R_2$ *for cables with copper and aluminium conductors, and this can be used to determine the maximum length of run for the circuit to comply with Regulation 413-4.*

PROTECTION AGAINST OVERCURRENT

As well as providing protection against electric shock, electrical installations must be protected from excess current (overcurrent) which may damage equipment or cause a fire. The term 'overcurrent' may be subdivided into two categories:-

- overload current
- short circuit current

OVERLOAD CURRENT

Overload currents usually occur because the equipment is overloaded, (drawing excess current), or the installation is abused, or has been badly designed, (a failure to correctly calculate the Diversity factor for example), or has been modified by an incompetent person. The danger in all such cases is that the temperature of the conductors will increase to such an extent that the effectiveness of any insulating materials will be impaired.

The devices used to detect such overload currents, and to break the circuit when they occur are the **FUSE**, (either the rewirable fuse or replaceable fuse commonly used with most domestic electrical equipment, or the HBC fuse) and the **MINIATURE CIRCUIT BREAKER** (MCB).

In order to protect against overload current the protective device must be rated greater than or at least equal to the design current; and the current carrying capacity of the cables must be greater than (or equal to) the rating of the protective device.

PROTECTION AGAINST SHORT-CIRCUIT CURRENTS

Whereas overload currents, as their name implies are likely to result in a current of perhaps no more than twice or three times the normal circuit current, a short–circuit current may be several hundred, or even several thousand times normal. In these circumstances the circuit protection must break the fault current **rapidly**, before danger is caused through overheating or mechanical stress.

The current *likely* to flow under short–circuit conditions is called the *prospective short-circuit current* (Ip), the value of which can be measured using a special test instrument, (where the circuit has been installed), or obtained from the supply authority.

If the device used for overload protection is also capable of breaking a prospective short circuit current (Ip) safely, it may be used for both overload and short circuit protection.

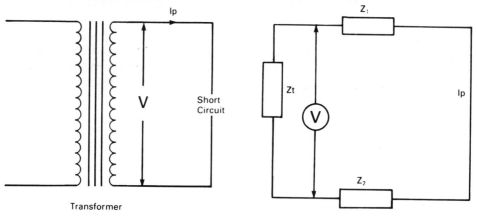

Short Circuit of negligible impedance.

Transformer

Overload protective devices can provide both overload and short-circuit protection if the device has a breaking capacity not less than the value of Ip and the following co-ordinating criteria are achieved.

$$I_B \le I_n \le I_Z$$

where:

I_B = design current of the circuit

I_n = nominal current or current setting of device

I_Z = current–carrying capacity of any of the circuit conductors.

PROTECTIVE DEVICES

FUSES

Types of fuses

Semi-enclosed referred to as rewireable (BS 3036)

Cartridge (BS 1361) and (BS 1362)

High breaking capacity referred to as HBC (BS 88)

Semi-Enclosed (Rewirable) Fuses (BS 3036)

Advantages of rewireable fuses

No mechanical moving parts

Cheap initial cost

Simple to observe whether element has melted

Low cost of replacing element

Disadvantages of rewireable fuses

Danger of insertion with fault on installation

Can be repaired with incorrect size fuse wire

Element cannot be replaced quickly

Deteriorate with age

Lack of discrimination

Can cause damage in conditions of severe short–circuit.

Where semi-enclosed (rewireable) fuses are used the fuse element must be fitted in accordance with manufacturers instructions, eg. correct c.s.a. and length of element.

In the absence of such instructions the fuse must be fitted with a single element of plain or tinned copper wire of the appropriate diameter as listed in the Table 53A of the IEE Wiring Regulations.

Table 53A

Nominal current of fuse (A)	Nominal diameter of wire (mm)
3	0.15
5	0.2
10	0.35
15	0.5
20	0.6
25	0.75
30	0.85
45	1.25
60	1.53
80	1.8
100	2.0

Cartridge Fuses (BS 1361, BS 1362)

The body of the fuse can be either ceramic (low grade) or glass with metal end caps to which the fuse element is connected. The fuse is sometimes filled with silica sand.

Advantages of cartridge fuses

Small physical size

No mechanical moving parts

Accurate current rating

Not liable to deterioration

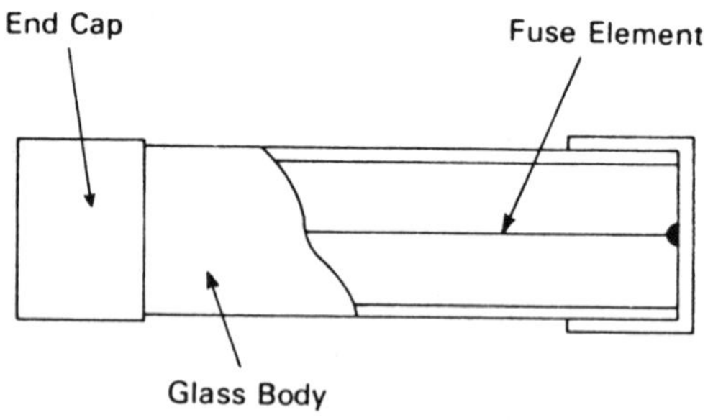

End Cap   Fuse Element

Glass Body

Disadvantages of cartridge fuses

More expensive to replace than rewireable fuse elements

Can be replaced with incorrect cartridge (mainly BS 1361 type)

Not suitable where extremely high fault current may develop

Can be shorted out by the use of silver foil

HBC Fuses (BS 88)

The barrel of the High Breaking Capacity fuse is made from high grade ceramic to withstand the mechanical forces of heavy current interruption.

Plated end caps afford good electrical contact.

An accurately machined element usually made of silver is shaped to give precise characteristic.

The fuse may be fitted with an indicator bead which shows when it has blown.

Advantages of HBC fuse

Consistent in operation

Reliable

Discriminates between overload currents of short duration, (eg. motor starting) and high fault currents.

Disadvantage of HBC fuse

Expensive

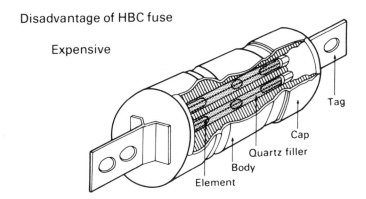

HBC fuse construction

The fuse-element consists of several parallel strips of pure silver with notches cut at pre-determined positions. A short length of pure tinfoil is wrapped round the centre notch of each strip.

The silver strips are spot-welded to silver-plated copper end rings.

This type of construction gives the required fusing factor as well as a time/current characteristic suitable for the apparatus to be protected. It also provides freedom from deterioration and limits the temperature rise. This is because the addition of pure tin to each silver strip of the fuse-element lowers the melting point of the combination and enables the fuse to operate at a fusing factor in the order of 1.4 without excessive temperature rise at the rated current.

The use of a composite silver/tin fuse-element permits the cross-sectional area to be much larger than that of a pure silver fuse-element, with the result that its thermal capacity is greater, giving a longer time-lag with comparatively small over-currents.

The notching of the fuse-element ensures that the circuit is cleared quickly with short-circuit currents.

The tin is wrapped round the silver cord so that the fuse-element works with all the freedom from deterioration of a pure silver fuse-element in ordinary service conditions. When an over-current occurs, the tin and the silver alloy melt, breaking the circuit.

Freedom from deterioration is further ensured by the spot-welding of the ends of the fuse-element.

The cartridge has a substantial ceramic core, with a silver-plated copper end ring secured to each end to form the end connections of the fuse-element. Electro-tinned brass end caps are pressed over the copper end rings and the cartridge is finally sealed by means of outer end rings of electro-tinned steel spun into grooves in the fuse-core. A fibre washer is interposed between the spun-on outer end ring and the brass end cap.

All cartridges are filled with silica sand to ensure quick and efficient arc-extinction in all conditions of operations.

Colour Coding of Fuses

BS 646

Rating (amps)	Colour
1	Green
2	Yellow
3	Black
5	Red

BS 1362

Rating (amps)	Colour
3	Red
13	Brown

BS 1361

Rating (amps)	Colour
5	White
15	Blue
20	Yellow
30	Red
45	Green
60	Purple

Note: The colour for all other ratings is black.

MINIATURE CIRCUIT BREAKERS (MCB's) (BS 3871)

British Standard

Miniature circuit breakers should be manufactured to BS 3871

Types of MCB:

Thermal and magnetic

Magnetic hydraulic

Assisted bimetal

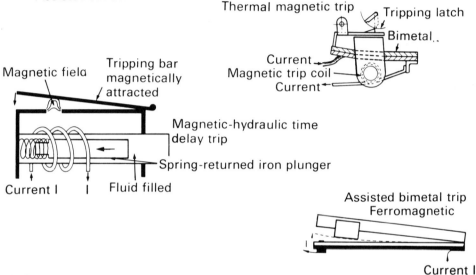

Thermal magnetic trip — Tripping latch — Bimetal — Current — Magnetic trip coil — Current

Magnetic field — Tripping bar magnetically attracted — Magnetic-hydraulic time delay trip — Spring-returned iron plunger — Current I — Fluid filled

Assisted bimetal trip Ferromagnetic — Current I

Advantages

Tripping characteristic set during manufacture; cannot be altered

They will trip for a sustained overload but not for transient overloads

Faulty circuit is easily identified

Supply quickly restored

Tamper proof

Multiple units available

Disadvantages

Have mechanically moving parts

Expensive

Need for regular testing to ensure satisfactory operation

Characteristics affected by ambient temperature

Thermal and Magnetic Tripping

In a thermal and magnetic type miniature circuit-breaker the time-delay effect in small and moderate overload conditions is provided by means of a thermally operated bimetal element.

When the bimetal element is heated to a pre-determined temperature, the resultant deflection is arranged to trip the circuit-breaker. The time taken to heat the element to this temperature, provides the necessary time-delay characteristic. However, the current at which a thermal circuit-breaker will operate will be affected by the ambient temperature.

The bimetal element may be arranged to carry line current and so be directly self-heated. Alternatively, on the lower current ratings in particular, indirect heating may be used. With indirect heated bimetals the response to overload conditions will obviously be rather more sluggish than with directly heated units, due to the time taken for the heat to transfer from the heater to the bimetal element. This, incidentally, is a point which may be of interest when selecting a suitable unit for a duty cycle which involves withstanding frequent transient surges of current.

Thermal tripping gives an inverse time-delay characteristic on small and moderate overloads. For operation under short circuit conditions an additional magnetic trip element is provided which takes the form of either an entirely separate instantaneous magnetic trip coil or uses the bimetal as a simple one turn coil with appropriate pole piece. The trip is arranged to come into operation at a suitable multiple of full load current depending on which characteristic type is required.

To provide a completely separate instantaneous tripping element is rather expensive in small competitively priced miniature circuit-breakers. A modern version of a combined thermal and magnetic trip is illustrated.

A   Bimetal alone tripping breaker

B   Bimetal and magnetic trip, combining to trip breaker. Here the two tripping elements are arranged to operate the same tripping latch.

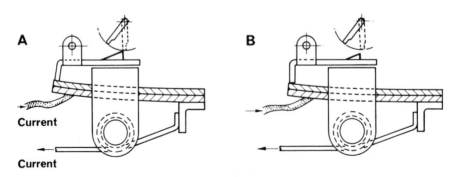

48

Magnetic trip alone tripping breaker (C)

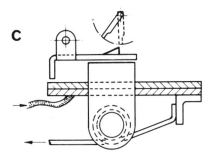

Magnetic/Hydraulic Tripping Mechanisms

The essential component of the magnetic/hydraulic time-delay tripping mechanism of this type of circuit-breaker is an hermetically sealed tube filled with silicone fluid and containing a closely fitted iron slug.

In normal load conditions, the magnetic pull from the trip coil is unable to over-come the restoring force of the time-delay spring and the iron slug remains at the far end of the tube.

When an overload occurs, the magnetic pull causes the slug to move through the tube, the speed of travel being controlled by the magnitude of the overload.

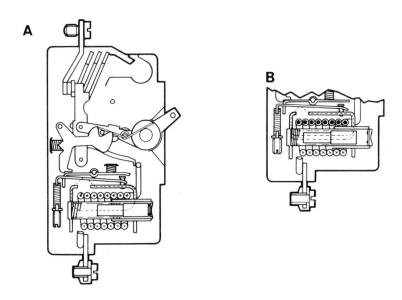

As the slug nears the end of the tube, the air gaps in the magnetic circuit are reduced, so increasing the magnetic pull on the armature to a point where it trips the circuit-breaker. This mechanism gives time-delay tripping under overload conditions, the time-delay being inversely proportional to the magnitude of the overload.

When a heavy overload or short-circuit occurs, the much greater magnetic pull on the armature is sufficient to trip the circuit-breaker instantaneously despite the gaps in the magnetic circuit; the slug does not have to move from its 'at rest' position. Thus the same mechanism gives time-delay tripping up to about seven times rated current and instantaneous tripping above that level.

Assisted Bimetal Tripping

In the assisted bimetal form of construction, the time-delay characteristic is again provided by a thermally operated bimetal element which may be either directly or indirectly heated. Instantaneous tripping in short circuit conditions is achieved by arranging for a powerful magnetic pull to deflect the bimetal.

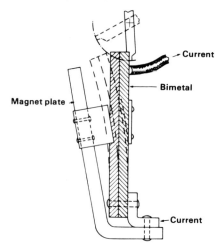

This method utilises the magnetic field which is produced when a current flows in a conductor. By locating the bimetal near to a substantial section of ferrous material, the magnetic field associated with current flowing in the bimetal will cause a sideways pull to be applied to the bimetal element, attracting the bimetal towards the ferrous material. This sideways pull is arranged to coincide in direction with the normal direction of movement of the bimetal.

In a small and moderate overload conditions the effect of this sideways pull may be negligible. In heavier overload conditions however, it becomes more significant. Initially, it simply helps to speed the bending of the bimetal. With a very heavy overload, as with a short-circuit, the sideways pull resulting from the magnetic field will in itself be powerful enough to deflect the bimetal sufficiently to trip the breaker.

Hence, with assisted bimetal tripping the miniature circuit-breaker will display a normal time-delay characteristic for small and moderate overloads. This delay will be reduced to instantaneous operation in short-circuit conditions. There is no clearly defined point at which instantaneous tripping takes place.

Residual Current Circuit-Breakers

Residual Current Circuit-Breakers give protection not only against fire risk but also give adequate protection against shock risk.

Basic Circuit

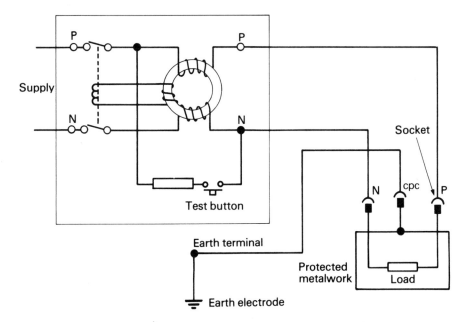

D.P. Switch: connected between incoming supply and load

Method of operation under fault conditions

The current taken by the load is fed through two equal and opposing coils wound on to a common transformer core. When the phase and neutral currents are balanced, (as they should be on a healthy circuit) they produce equal and opposing fluxes in the transformer core resulting in no voltage in the trip coil. If more current flows in the phase side than in the neutral side, an out-of-balance flux will be produced which will be detected by the fault detector coil. The fault detector coil opens the DP switch by energising the trip coil.

Test switch

The test switch is a requirement of BS 842. The test switch tests only that the circuit-breaker is functioning correctly and is operating in the correct order of sensitivity, as specified by BS 4293.

SELECTION OF LIVE CONDUCTORS

CURRENT CARRYING CAPACITIES OF CABLES

The rating of a cable depends on its ability to dissipate the heat generated by the current it carries. This in turn depends in part on the type of installation. Table 9A (Appendix 9) of the IEE Wiring Regulations lists twenty standard methods of installation with examples identified by numbers. These classifications are used in the current carrying capacity tables.

Example: From Table 9D2, it can be seen that ordinary twin and multicore PVC-insulated cable installed in an enclosure such as trunking has a current capacity of only approximately 85% of what it would be when clipped to a surface or embedded directly in plaster.

Factors which effect the ability of a cable to lose heat (other than its physical characteristics) are:

- ambient (surrounding air) temperature
- cable grouping
- thermal insulation

Ambient Temperature

The rate of heat loss from a cable depends on the difference in temperature between the cable and the surrounding air. A correction must be made where the cable is to be installed in high or low ambient temperatures.

Tables 9C1 and 9C2 of the IEE Wiring Regulations give correction factors to be applied to the tabulated current-carrying capacities depending upon the actual ambient temperature of the location in which the cable is to be installed.

Cable Grouping

Cables installed in the same enclosure and all carrying current will all get warm. Those near to the edge of the enclosure will be able to transmit heat outwards but will be restricted in losing heat inwards, while cables in the centre may find it difficult to lose heat at all.

The correction factors for this effect are given in Table 9B. These relate to 'touching' and 'spaced' cables.

It may well be that for a particular circuit the circumstances may change throughout its length, i.e. the ambient temperature or the number of cables bunched together may vary, or there may even be a change in the method of installation. Where this is the case the most onerous figure, i.e. that giving the LOWEST current carrying capacity must be used. Alternatively the correction factors may be applied to the length of run specifically affected, but this will require the size of the cable to be increased for this length.

Notes
1. *The factors in Table 9B relate to groups of cables, all of one size.*

2. *If due to known operating conditions a cable is expected to carry a current not more than 30% of its grouped rating it may be ignored when calculating the rating factor for the rest of the group.*

3. *Where spacing between adjacent cables exceeds twice the overall diameter, no reduction factor need be applied.*

Thermal Insulation

The increasing use of thermal insulation especially in roof spaces has led to the inclusion of Regulation (522-6). Thermal insulation is designed to limit heat flow, so a cable in contact with it will tend to become warmer than one not in contact. The Regulations therefore give correction factors to be applied where a significant length of the cable is in contact (or likely to be in contact) with such insulation.

Correction factor applicable when cable is **totally** surrounded = 0.5

This clearly means that any cable installed in a roof space where it has proved impossible to route the cable away from thermal insulation will only be able to carry half its rated current if completely surrounded.

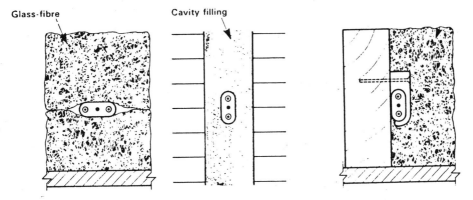

Glass-fibre   Cavity filling

When a cable is installed or likely to be installed in a thermally insulated wall or ceiling, but is in contact with the thermally conductive surface on *one side only*, the current-carrying capacity should be chosen from the appropriate Tables 9D1 to 9L1 in Appendix 9 of the IEE Wiring Regulations.

DETERMINATION OF THE SIZE OF CABLES

Having established the design current of a circuit and selected the type and size of protective device, it is necessary to determine the size of cable to be used.

Procedure

For single circuits

- Divide the nominal current of the protective device (I_n) by any applicable correction factor for ambient temperature (C_a) given in Table 9C1.
- Apply correction factor for thermal insulation (C_i) as a divisor
- Select the size of cable from the tables so that the current carrying capacity (I_t) is not less than the value of the protective device.

For groups of cables

- Divide the nominal current of the protective device (I_n) by the applicable correction factor for grouping (C_g) given in Table 9B.
- Apply correction factors for any applicable ambient temperature (C_a) from Table 9C1 for the type of insulation
- Apply correction factor for thermal insulation (C_i) if the cable is in contact with thermal insulation

When the protective device is other than a semi-enclosed fuse to BS 3036 the cable selected has to be such that its tabulated current carrying capacity for the method of installation is not less than

$$I_t \geqslant \frac{I_n}{C_g \times C_a \times C_i}$$

If the protective device is a semi-enclosed fuse to BS 3036

Where the protective device is a semi-enclosed fuse to BS 3036 in both single circuits and for groups of cables a further correction factor of 0.725 is to be applied (not applicable to MI cable installation) in these cases

$$I_t \geqslant \frac{I_n}{C_g \times C_a \times C_i \times 0.725}$$

For single circuits

- The nominal current of the protective device should be divided by the correction factors for ambient temperature given in Table 9B2
- Apply correction factor for thermal insulation as a divisor
- Then further divide by 0.725

  The size of the cable selected should be such that its tabulated current carrying is not less than the value of the protective device

For groups of cables

- Divide the nominal current of the protective device (I_n) by the applicable correction factor for grouping given in Table 9B1.

$$I_t \geqslant \frac{I_n}{0.725.F}$$

Where I_t is the required tabulated single-circuit current carrying capacity and F is the group rating factor

- Apply correction factors for ambient temperature from Table 9B2 as a divisor to the value of I_t
- Apply correction factor for thermal insulation as a divisor to the value of I_t
- Select the size of cable from the tables so that the current carrying capacity for the installation method is not less than the value of the protective device

Example

A 240V lighting circuit consisting of 10 × 100 watt tungsten lamps is wired in PVC insulated single-core cable with copper conductors. It is protected by a 5A BS 3036 fuse. The cable is run through an ambient temperature of 35°C and is grouped with two other lighting circuits, e.g. two phase and two neutral conductors which are of the same size, equally loaded and which are installed in the same conduit system on a wall. Determine the minimum cable size for compliance with the Regulation (522-1). The length of circuit run and voltage drop are to be neglected in this example.

Step 1

First check that the overcurrent protective device is adequate

$$\text{Design Current } I_b = \frac{\text{Power (W)}}{\text{Voltage (V)}}$$

$$= \frac{10 \times 100}{240} = \frac{1000}{240}$$

$$= 4.16A$$

Step 2

Correction factor C_g (Table 9B) for three circuits = 0.70

$$I_t \geqslant \frac{I_n}{0.725 \times C_g}$$

$$\geqslant \frac{5}{0.725 \times 0.7}$$

$$\geqslant 9.84A$$

Step 3

Determine correction factor (C_a) for the ambient temperature of 35°C (Table 9C2).

At 35°C correction factor with BS 3036 protective device = 0.97.

Step 4

Determine minimum current carrying capacity of circuit live conductors

$$I_t \geqslant \frac{9.84}{0.97} \; A$$

$$\geqslant 10.15A$$

Step 5

Select cable size from Appendix Table 9D1 (Installation method 3)

From column 4 1.0 mm^2 = 13.5A current carrying capacity of conductor.

Voltage Drop

A further consideration when selecting cable sizes is that of voltage drop. Regulation 522-8 requires that every bare conductor or cable should be selected so that the voltage drop within the installation should not exceed a value which might impair the safe function of any connected equipment, in normal service.

For final circuits protected by an overcurrent device, having a nominal current of not more than 100A. This requirement is satisfied if the voltage drop from the point of origin to any other point in the circuit does not exceed 2.5% of the nominal voltage, disregarding starting conditions.

Current carrying capacity tables (Appendix 9) include figures from which the cable volt drop can be calculated. Each size of every cable type is given a figure expressed as voltage drop per ampere per metre (in millivolts).

To calculate the volt drop, this figure must be multiplied by the length of the cable (in metres) and the current on full load. (The final product must be divided by 1000 to give the answer in volts).

Diversity must be taken into account when calculating volt drop. The application of rating factors means that in many cases the actual current is much less than the rated current, and the cable is cooler, and thus has a lower resistance than that calculated.

Calculation based on the tables will give a value for volt drop that is the highest possible figure. In marginal cases, an exact calculation, based on actual conductor resistance may be preferred.

Example

A 240V, 30A single-phase circuit consists of an 18m length of run of PVC insulated cable installed in conduit. The circuit has a full load current of 26A. Determine the minimum size of cable which will comply with voltage drop requirements of Regulation 522-8.

$$\text{Maximum voltage drop allowed} = \frac{240 \times 2.5}{100} = 6V$$

Actual voltage drop

$$= \frac{mV/A/m \times \text{Design current} \times \text{length}}{1000}$$

The simplest method to adopt is to determine the *maximum* mV/A/m which will comply with the *maximum voltage drop allowed,* in this case 6V.

$$\text{i.e. } 6V = \frac{mV/A/m \times 26 \times 18}{1000}$$

$$\text{max } mV/A/m = \frac{6000}{26 \times 18}$$

$$= 12.8 \ mV/A/m$$

By referring to Table 9D1, *any* cable with 12.8 mV/A/m *or less* will therefore give an actual voltage drop of 6V *or less.*

Table 9D1 shows:

2.5 mm² = 18mV/A/m (hence v.d. *greater* than 6V)

$$\text{i.e. } \frac{18 \times 26 \times 18}{1000} = 8.42V$$

4.0 mm² = 11 mV/A/m (hence v.d. **less** than 6V)

i.e. $\dfrac{11 \times 26 \times 18}{1000}$ = 5.15V

The minimum size of cable which may be selected in this case is therefore 4.0 mm².

Note: The above cable selection shows compliance only with Regulation 522-8.

SIZING CONDUIT AND TRUNKING SYSTEMS

SPACE FACTORS

The number of cables which can be drawn in or laid in any enclosure of a wiring system must be such that no damage can occur to the cables or the enclosure during installation. In order to achieve this, conduit and trunking systems are given a 'space factor'.

In order to comply with the above requirement a method employing a 'unit system' is described in Appendix 12 of the Regulations, where each cable is allocated a factor. The sum of the factors for the cables which are to be run in the same enclosure is then compared with a factor given in the tables for different sizes of conduit or trunking, in order to determine the size of conduit or trunking necessary.

The Tables in Appendix 12 of the Regulations cover the following parts of the 'unit system'.

Wiring System

- Straight runs of conduit not exceeding 3m (Table 12B)
- Straight runs of conduit exceeding 3m or run of any length incorporating bends or sets (Table 12D)
- Trunking (Table 12F)

Note: For conduit systems a bend is classed as 90° and a double set is equivalent to one bend.

Cables

- Single core PVC insulated cables in straight runs of conduit not exceeding 3m in length (Table 12A)
- Single core PVC insulated cables in straight runs of conduit exceeding 3m in length or runs of any length incorporating bends and sets (Table 12C)
- Single core PVC insulated cables in trunking (Table 12E)
- Other sizes and types of cable in trunking; the space factor should not exceed 45%

Note: Only mechanical considerations have been taken into account in determining the factors in the tables. The values for cable capacities given in the Appendix have been based on the application of an easy pull of cables into conduits.

As the number of cables increases in a conduit or trunking the current carrying capacity of the cables must be reduced in accordance with the application of the grouping factors in Appendix 9. In such cases it may be more economical to separate the circuits and run the conductors in more than one enclosure.

Example

A lighting circuit for a village hall requires the installation of a conduit system with a conduit run of 10m with two right angle bends; the number of cables required is ten 1.5mm² PVC insulated. What size conduit should be chosen for the installation?

Step 1

Select correct table for cable runs over 3m with bends (Table 12C)

Step 2

Obtain factor for 1.5mm² cable – 22

Step 3

Apply factor to number of cables in run = 22 × 10 = 220

Step 4

Select correct table for conduit systems with run in excess of 3m with bends (Table 12D)

Step 5

Obtain from table a factor for the length of run which is greater than 220. The table gives a factor of 260 for a 10 metre run in conduit with two bends – 260.

Step 6

From the table the conduit size required is 25mm – *Answer* conduit is 25mm.

When other sizes and types of cable or trunking are used the space factor should not exceed 45%. In this situation it is necessary to carry out the following procedure to determine the size of trunking required after the cable size has been decided.

- Consult the cable manufacturer's literature to obtain the overall dimensions of the cable including the insulation.

- Work out the cross sectional areas of the cables which are to be installed by using the formula:

  Cross sectional area $= \dfrac{\pi \times d^2}{4}$

- Add together the individual cross sectional areas of the cables concerned and obtain the total cross sectional area of the cables.

- Obtain the size of trunking by using the following formula which will allow a 45% space factor.

  let A be the c.s.a. of the trunking required

  $A \times \dfrac{45}{100}$ mm^2 = total c.s.a. of the cables

  then $A = \dfrac{100}{45} \times$ total c.s.a. of cables

- Check manufacturer's trunking sizes and select one size. Convert to a c.s.a. and compare with calculated value.

Example

A steel trunking is to be installed as the wiring system for 8 single-phase circuits each having a design current of 35A.

BS 88 Part 2 fuses (40A) will be used as the overcurrent protective devices and PVC-insulated copper cables will be installed.

Determine the size of trunking required.

Step 1

C_g is the grouping factor and from Table 9B, the value of eight circuits in trunking = 0.52.

Minimum current carrying capacity of cables

$$\therefore I_T = \frac{I_n}{C_g} = \frac{40}{0.52} = 76.92A$$

Using Table 9D1 and installation method B the minimum size which can be used is 25mm² cable.

Step 2

For 25mm² cable the overall c.s.a. is 63.8mm² (obtained from manufacturer's data).

Step 3

$$\text{Total c.s.a. of cables} = 16 \times 63.8$$

$$= 1021mm^2$$

Step 4

$$\text{Minimum trunking c.s.a. (A)} = \frac{100}{45} \times \text{total c.s.a. of cables}$$

$$\text{(A)} = \frac{100}{45} \times 1021$$

$$= 2269mm^2$$

Step 5

From manufacturer's data the nearest trunking size greater than A is 50mm × 50 mm (2500 mm²)

EARTHING AND PROTECTIVE CONDUCTORS

EARTHING

The earth can be considered to be a large conductor which is at zero potential. The purpose of earthing is to connect together all metal work (other than that which is intended to carry current) to earth so that dangerous potential differences cannot exist either between different metal parts, or between metal parts and earth.

Purpose of earthing

By connecting to earth metalwork not intended to carry current, a path is provided for leakage current which can be detected and interrupted by the following devices:-

- fuses
- circuit breakers
- residual current devices

Connections to earth

The earthing arrangement of an installation must be such that:-

- The value of resistance from the consumer's main earthing terminal to the point of earthing is in accordance with the protective and functional requirements of the installation and expected to remain continuously effective.

- Earth fault and earth leakage currents which may occur under fault conditions can be carried without danger.

- They are robust and protected from mechanical damage.

EARTHING OF SUPPLYING SYSTEMS

TN-S system

This is likely to be the type of system used where the Electricity Board's installation is fed from underground cables with metal sheaths and armour. In TN-S systems the consumer's earthing terminal is connected by the supply authority to their protective conductor (i.e. the metal sheath and armour of the underground cable network) which provides a continuous path back to the star point of the supply transformer, which is effectively connected to earth.

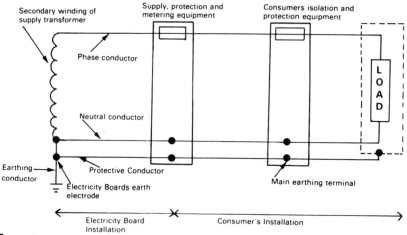

TT systems

This is likely to be the installation used where the Electricity Board's installation is fed from overhead cables, where no earth terminal is supplied. With such systems the earth electrode for connecting the circuit protective conductors to earth has to be provided by the consumer. Effective earth connection is sometimes difficult and in such cases a residual current device should be installed.

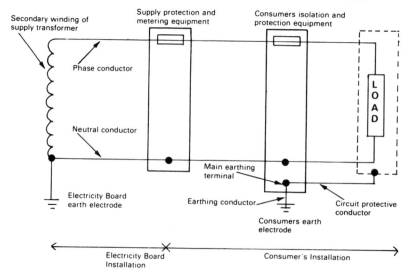

TN-C-S systems

Where the Electricity Board installation uses a combined protective and neutral (PEN) conductor, this is known as a TN-C supply system. Where consumer's installations consisting of separate neutral and earth (TN-S) are connected to the TN-C supply system, this combination is called a TN-C-S system.

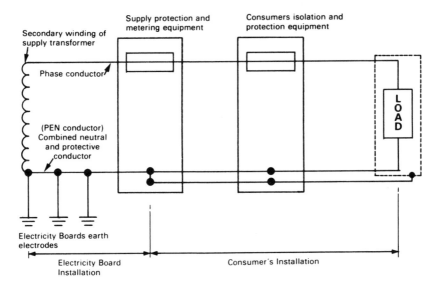

General

In the United Kingdom the Electricity Boards have to comply with the 1937 Electricity Supply Regulations which require that the supply network shall be directly earthed at one point, the requirement being modified when appropriate by the PME Approval 1974. This approval permits the Electricity Boards to earth the supply network at more than one point; therefore the majority of systems in use in the United Kingdom are TN-S, TN-C-S and TT systems.

Full discussions with the supply authority are essential when planning or installing a 'consumers' system for use on a PME supply.

TYPES OF PROTECTIVE CONDUCTORS

- PVC insulated single core cable manufactured to BS 6004, colour green/yellow

- PVC insulated and sheathed cable with an integral protective conductor manufactured to BS 6004

- Copper strip

- Metal conduit ⎤
- Metal trunking systems ⎬ enclosures
- Metal ducting ⎦
- MICC - cable sheath
- Lead covered cable sheath
- SWA cable armourings

When the protective conductor is formed by a wiring system such as conduit, trunking, MICC or armoured cables, a separate protective conductor must be installed from the earthing terminal of socket outlets to the earthing terminal of the associated box or enclosure.

The circuit protective conductor of final ring circuits which are not formed by the metal covering or enclosures of a cable should be installed in the form of a ring having both ends connected to the earth terminal at the origin of the circuit, e.g. distribution board or consumer's unit.

Earthing Arrangements and Network Terminations Relating to the Various Systems

N.B. Systems IT and TN-C are not
envisaged for general use and are,
therefore, not shown.

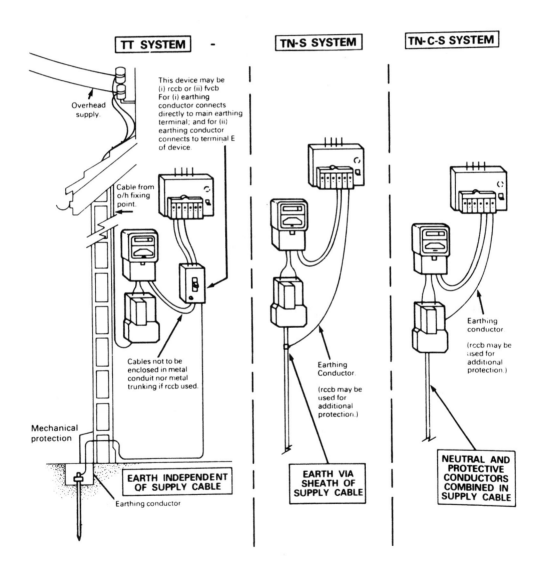

TT SYSTEM -

This device may be
(i) rccb or (ii) fvcb
For (i) earthing
conductor connects
directly to main earthing
terminal; and for (ii)
earthing conductor
connects to terminal E
of device.

Overhead
supply.

Cable from
o/h fixing
point.

Cables not to be
enclosed in metal
conduit nor metal
trunking if rccb used.

Mechanical
protection

**EARTH INDEPENDENT
OF SUPPLY CABLE**

Earthing conductor

TN-S SYSTEM

Earthing
Conductor.

(rccb may be
used for
additional
protection.)

**EARTH VIA
SHEATH OF
SUPPLY CABLE**

TN-C-S SYSTEM

Earthing
conductor

(rccb may be
used for
additional
protection.)

**NEUTRAL AND
PROTECTIVE
CONDUCTORS
COMBINED IN
SUPPLY CABLE**

Flexible conduit cannot be used as a protective conductor; an additional circuit protective conductor should be installed.

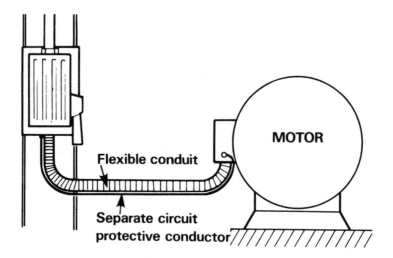

Flexible conduit

MOTOR

Separate circuit protective conductor

Main earthing terminal bars

A main earthing terminal or bar must be provided in an accessible position for every installation, for the connection of the circuit protective conductors and main bonding conductors to the earthing conductor.

The method of disconnecting the earthing terminal from the means of earthing must be such that it requires the use of tools such as spanners or screwdrivers.

MAIN EQUIPOTENTIAL BONDING

Equipotential bonding to gas and water services should be made at their point of entry using bonding conductors with cross-sectional areas of not less than half the cross-sectional area of the earthing conductor; the minimum size being 6mm². Except where PME conditions apply, the cross-sectional area need not exceed 25mm² if the bonding conductor is of copper; or equivalent conductance where other metals are used.

Note: Supply authorities should always be consulted where PME installations are provided to check if any special requirements exist.

Bonding of gas service pipes should be made on the consumer's side of the meter between the meter outlet union and any branch pipework, but within 600mm of the gas meter as illustrated.

The purpose of installing bonding conductors is to ensure that any metalwork within an installation, such as gas and water services, are at the same potential as the metalwork of the electrical installation.

Note: *Check with local electricity, gas and water authorities for any special requirements regarding the bonding of services.*

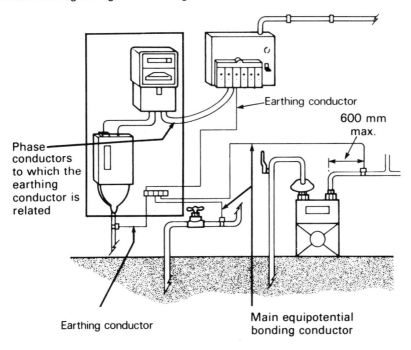

Earthing conductor

600 mm max.

Phase conductors to which the earthing conductor is related

Earthing conductor

Main equipotential bonding conductor

Supplementary bonding

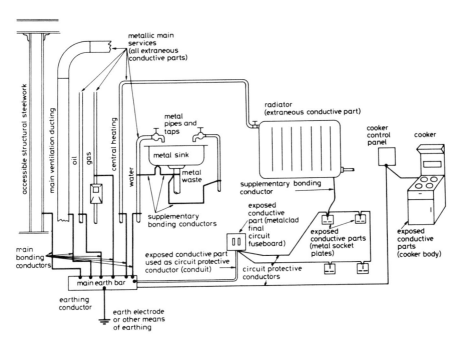

A supplementary bonding conductor used to connect exposed conductive parts together, or exposed conductive parts to extraneous conductive parts must have a cross-sectional area not less than the smallest protective conductor used to make the connection to the exposed conductive part, subject to a minimum of:-

- 2.5mm^2 if mechanically protected

- 4mm^2 if mechanical protection is not provided

In situations where a supplementary bonding conductor connects two extraneous conductive parts, neither of which are connected to an exposed conductive part, the minimum cross-sectional area of the supplementary bonding conductor shall be:-

- 2.5mm^2 if mechanically protected

- 4mm^2 if mechanical protection is not provided

Supplementary bonding conductors may need to be installed in situations such as kitchens where a person may make simultaneous contact with an electrical appliance (such as an electric kettle or washing machine) and other metalwork, (such as the hot or cold water taps). In these situations supplementary bonding may be required as illustrated (but not if earth continuity tests prove that all metalwork of electrical, gas and water services its effectively bonded). The pipework of the gas and water services may be effectively connected together using permanent and reliable metal to metal joints of negligible impedance, such as Yorkshire fittings.

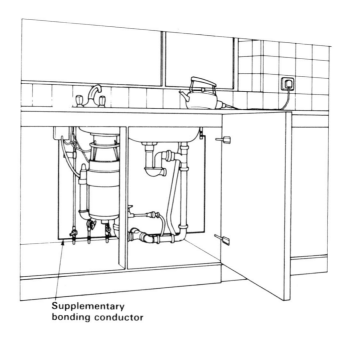

Supplementary
bonding conductor

In rooms containing a fixed bath or shower supplementary bonding conductors should be installed, to reduce to a minimum the risk of electric shock in circumstances when the body resistance is likely to be low. The most common method of making a bonding connection to pipework is by using BS 951 earth clips.

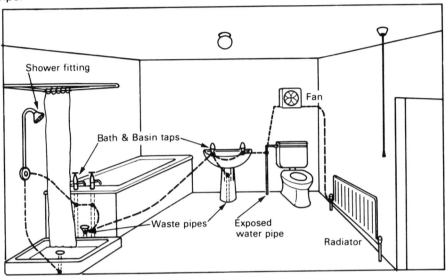

EARTHING CLIPS

Earth clips are commonly used for primary equipotential bonding. Specially manufactured clips are designed to make a secure and sound electrical contact with pipes carrying main services. The method to be adopted should be as follows:-

- select size of clip to suit the size of pipe to be bonded
- clean pipe using emery cloth or wire wool
- place strap round pipe and through the clip and tighten using a screwdriver; tighten down locking nut
- fit warning label to the bonding conductor
- strip the insulation from the bonding conductor and connect to terminal on clip and tighten; avoid cutting the bonding conductor when looping to another service pipe
- check that the earthing clip and bonding conductor are tight

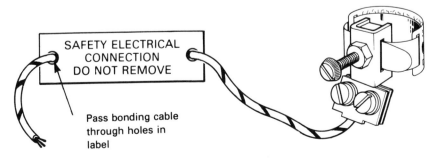

Termination to Earth Electrode

The connection of earthing conductors to electrodes require adequate insulation where they enter the ground, to avoid possible dangerous voltage gradient at the surface. All electrode connections must be thoroughly protected against corrosion and mechanical failure.

It is important that the electrode is made accessible for inspection purposes, and a label should be fitted at or near the point of connection.

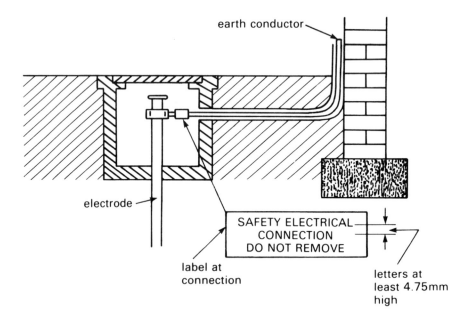

earth conductor

electrode

label at connection

SAFETY ELECTRICAL
CONNECTION
DO NOT REMOVE

letters at least 4.75mm high

CABLE SELECTION

SELECTING CABLES FOR CIRCUITS

When installing a circuit, it is necessary to:

- – Calculate the design current (I_b)
- – Select the type and nominal rating of the protective device (I_n)
- – Determine and apply correction factors to I_n
- – Select cable from tables in Appendix 9 (I_z)
- – Calculate the voltage drop and check for compliance
- – Check that circuit complies with shock protection
- – Check that circuit complies with thermal constraints

Example 1

A 20A radial socket outlet circuit is protected by a BS 88 fuse. The circuit is wired using 2.5 mm^2 single core PVC cables installed in a 16 m length of PVC conduit. A separate protective conductor consisting of a 1.0 mm^2 PVC cable is used. Assuming that no rating factors are applicable and that the value of Z_E is given as 0.5 ohms, determine whether the circuit complies with the IEE Regulations. The nominal voltage (U_o) may be taken as 240V.

    I_b Design current of circuit = 20A

    I_n Protective device, BS 88 fuse = 20A

Step 1

    Apply correction factors for grouping and ambient temperature.

        None apply

Step 2

    Select suitable cable (I_z) from Appendix 9

    from Table 9D1 2.5 mm^2 cable = 24A

Step 3

Calculate voltage drop

Maximum volt drop allowed = 2.5% of 240 = 6V

Actual voltage drop =

$$\frac{mV/A/m \times \text{design current Ib} \times \text{length}}{1000}$$

Actual v.d. = $\dfrac{18 \times 20 \times 16}{1000}$

= 5.76V

Step 4

Check for compliance with shock and thermal protection

Total circuit impedance = Z_s

Maximum Z_s from Table 8B (i) = 1.8 ohms

Actual Z_s = Z_E + (R_1 + R_2 ohms/metre × length)
Z_E = 0.5 ohms

R_1 + R_2 ohms/metre = 25.51 milli ohms per metre (from Table 17A)

Convert to ohms/metre = $\dfrac{25.51}{1000}$ × 16 (length of cable run)

= 0.408

Apply multiplication factor from Table 17B

Actual Z_s = 0.5 + (0.408 × 1.38)

= 0.5 + 0.56

Actual Z_s = 1.06 *(satisfactory)*

The actual Z_s value is less than the maximum Z_s value, therefore the live conductors and the cpc are suitable and the circuit complies with the Regulations.

Example 2

A 240v, 14kW domestic electric cooker installed in a house is to be supplied by a 15m run of PVC insulated and sheathed cable clipped to a surface. The cooker control unit incorporates a 13A socket outlet. The cable is run for a considerable part of its length in a roof space where it is covered on one side by thermal insulation. The circuit is to be protected by a BS 3036 fuse and the value of external impedance is 0.85 ohms. Determine the minimum size of cable which may be used.

Step 1

Determine full load current

$$I = \frac{W}{V} = \frac{14,000}{240}$$

$$= 58.33A$$

Step 2

Determine design current Ib

Apply allowance for diversity from Table 4B Appendix 4

Diversity for cooker = 10A + 30% of remaining current + 5 A

$$\text{design current Ib} = 10 + (\frac{30 \times 48.33}{100}) + 5$$

$$= 29.5A$$

Step 3

Protective device In chosen

$$= 30A \text{ BS } 3036$$

Step 4

Select cable size

$$I_t \geqslant \frac{I_n}{0.725} \quad \text{(BS 3036 fuse protection factor = 0.725)}$$

$$\geqslant 41.37 \text{ A}$$

Select from Table 9D2, Reference Method 3

$$10mm^2 \text{ size} = 52A$$

Cable size to be used is 10mm^2 phase with 4mm^2 cpc

Step 5

Calculate voltage drop

Maximum v.d. = 2.5% of 240V

= 6V

Actual voltage drop = $\dfrac{\text{mV/A/m} \times \text{Ib} \times \text{length}}{1000}$

actual voltage drop = $\dfrac{4.4 \times 29.5 \times 15}{1000}$

= 1.95V *(satisfactory)*

Step 6

Check for shock protection constraints

from Table 41 A1, Z_s max = 1.1Ω

(used because of socket outlet in cooker control unit)

actual Z_s = Z_E + $(R_1 + R_2)$ ohms

Z_E = 0.85Ω

$(R_1 + R_2)$ = 6.44 milliohm/metre (from Table 17A)

(Assuming 10 mm² phase conductor/4mm² cpc)

Multiplier (for (PVC)

= 1.38

Actual Z_s = 0.85 + $\dfrac{(6.44 \times 1.38 \times 15)}{1000}$

= 0.98Ω

Check: actual Z_s value is less than the maximum Z_s value, therefore 10mm²
with 4mm² c.p.c. cable is satisfactory.

INSPECTION AND TESTING

REASONS FOR INSPECTION AND TESTING

The purpose of inspection and testing of electrical installations is to verify that installations are safe and comply with the requirements of Regulations.

GENERAL

Every completed installation must be inspected and tested before being connected to the supply and energised. This should be done in such a manner that no danger to persons or damage to property or equipment can occur, even if the circuit tested is defective.

The following information should be made available to the persons carrying out the inspection and testing of an installation.

- Diagrams, charts or tables indicating:-

  (a)   the type of circuits,
  (b)   the number of points installed,
  (c)   the number and size of conductor,
  (d)   the type of wiring system.

- The location and types of devices used for:-

  – protection
  – isolation and switching

- Details of the protection devices, the earthing arrangements for the installation, the impedances of the circuits.

Note: Regulations 514-3 permits information for simple installations to be given in a schedule. Table 16 gives an example for a domestic installation. Where fixed equipment is installed, as in a bathroom, additional information should be given.

TABLE 16

Schedule of installation at ....
as prescribed in the IEE 'Regulations for electrical installations'

Type of circuit	Points served	Phase Conductor mm²	Protective Conductor mm²	Protective and switching devices
Ring final circuit	General purpose 13A socket outlets	2.5	1.5	30A miniature circuit breaker; type 2 local switches and plugs and socket outlets
No.1 lighting circuit	Downstairs fixed lighting	1.5	1.0	5A mcb; type 2 local switches
No. 2 lighting circuit	Upstairs fixed lighting	1.5	1.0	5A mcb; type 2 local switches
Cooker circuit	Cooker	6.0	2.5	45A mcb; type 2 cooker control unit
Immersion heater circuit	Immersion heater	2.5	1.5	15A mcb; type 2 local switched connector box

Type of wiring: PVC insulated and sheathed, flat twin with cpc BS 6004.

INITIAL INSPECTION

Visual Inspection

A visual inspection should be made of an installation to verify that installed electrical equipment

- complies with the appropriate British Standards (this may be ascertained by mark or by certificate furnished by the installer or manufacturer)

- is correctly selected and erected in accordance with these Regulations

- is not visibly damaged so as to impair safety.

CHECK LIST - Initial Inspection

- Connection of conductors

- Identification of conductors

- Selection of conductors for current-carrying capacity and voltage drop

- Connection of single-pole devices for protection or switching in phase conductors only

- Correct connection of socket outlets and lampholders

- Presence of fire barriers and protection against thermal effects

- Methods of protection against direct contact (including measurements of distances where appropriate) ie.

    - protection by insulation of live parts
    - protection by barriers or enclosures
    - protection by obstacles
    - protection by placing out of reach

- Protection by non-conducting location

- Presence of appropriate devices for isolation and switching

- Choice and setting of protective and monitoring devices

- Labelling of circuits, fuses, switches and terminals

- Selection of equipment and protective measures appropriate to external influences

- Presence of danger notices and other warning notices

- Presence of diagrams, instructions and similar information

Note: During any re-inspection of an installation all pertinent items in the check list should be covered.

TESTING

Test Instruments

Test instruments should be regularly checked and re-calibrated to ensure accuracy. The serial number of the instrument used should be recorded with test results, to avoid unnecessary re-testing if one of a number of instruments is found to be inaccurate.

For operation, use and care of test instruments, refer to manufacturers handbook.

The following items, (where relevant to the installation being tested), must be tested in the following sequence

- Continuity of ring final circuit conductors*

- Continuity of protective conductors including main and supplementary bonding*

- Earth electrode resistance

- Insulation resistance*

- Insulation of site-built assemblies

- Protection by electrical separation

- Protection by barriers or enclosures provided during erection

- Insulation of non-conducting floors and walls

- Polarity*

- Earth fault loop impedance

- Operation of residual current and fault–voltage operated protective devices.

*Note: Only items marked with * are dealt with in this edition of Study Notes.*

Standard methods of testing are described in Appendix 15 of the IEE Wiring Regulations. The use of other methods of testing is not precluded provided that they give results which are not less effective.

If a test indicates failure to comply, that test, and the preceding tests (whose results may have been affected by the fault) must be repeated after rectification of the fault.

CONTINUITY OF RING FINAL CIRCUIT CONDUCTORS

A test must be made to verify the continuity of all line and protective conductors to every final ring circuit.

Two methods of testing are indicated in the IEE Regulations. Both methods assume that an outlet is installed near the mid-point circuit. Where this is the case, these tests effectively establish that the ring has not been interconnected to create an apparently continuous ring circuit where an actual break exists as illustrated. Only the simpler method is described, below.

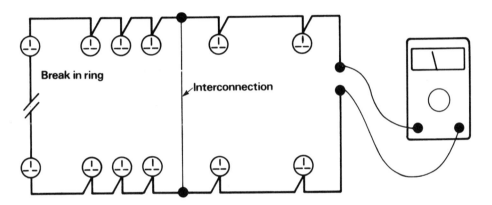

One core illustrated for clarity.

As it is unlikely that the socket outlet nearest the mid-point of a ring circuit is labelled, it can be located by visual inspection. It must be remembered that the socket outlet nearest the mid-point may well be the final socket outlet, as illustrated below.

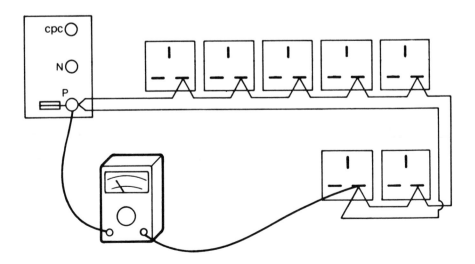

Method

Step A

The continuity of the phase conductor, neutral conductor and protective conductor is measured at the distribution board between the ends of each of the conductors before completion of the ring using a continuity tester. The resistance readings for each conductor is recorded.

Step B

Measure the resistance of the long test leads (used in Step C) and note the value of resistance.

Step C

After reconnection or initial connection the resistance is measured between the phase conductor terminal in the distribution board and the appropriate terminal or contact at the socket outlet or point nearest to the mid-point of the ring. The resistance value of the phase conductor is recorded.

The resistance values of test leads is deducted from the value measured in Step B. The resulting value should be approximately one quarter of the corresponding value obtained for the test between the ends of the phase conductor taken at the distribution board (Step A). The same test is carried out on the neutral conductors and protective conductors.

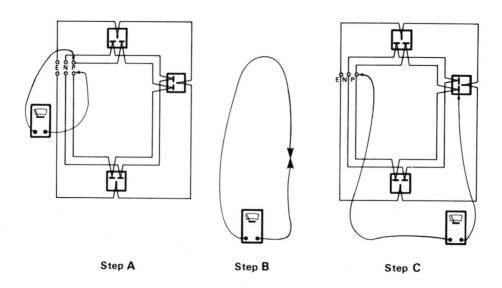

Step A   Step B   Step C

Example

Assume the results of the tests on the phase conductor were

A   test between ends of ring at distribution board 0.4Ω (ignoring resistance of short lead)

B   resistance of long test lead 0.1Ω

C   test from closed ring terminal at the distribution board to centre of ring (via long lead) 0.2 Ω

Therefore resistance of closed ring circuit conductor to centre point

$$= \ 0.2 \ - 0.1 \ = 0.1Ω$$

To verify the reading of the phase conductor to the centre of ring is about one quarter of reading taken between phase conductors at distribution board

Check $\dfrac{0.4}{4}$ = 0.1Ω *Test result satisfactory*

Summary

● Measure resistance between ends of phase conductor (A)

● Measure resistance of test lead (B)

● Measure resistance from closed ends of phase conductor to mid-point of ring (C)

● Check that reading $\dfrac{A}{4}$ = C - B

● Repeat tests for neutral and protective conductors

Visual inspection of ring circuit conductors

An alternative to the above methods for verifying that no interconnection multiple loops have been made in a ring circuit is for the installer to inspect each conductor throughout its entire length.

CONTINUITY OF PROTECTIVE CONDUCTORS

The initial tests applied to protective conductors are intended to verify that the conductors are both correctly connected and electrically sound, and also the resistance is such that the overall earth fault loop impedance of the circuits is of a suitable value to allow the circuit to be disconnected from the supply in the event of an earth fault.

Tests may be made using an a.c. or d.c. source of supply not exceeding 50 volts and with a current approaching 1.5 times the design current of the circuit, (but need not exceed 25 A).

When an a.c. test is used the current should be at the supply frequency.

When d.c. test is used the protective conductor should be inspected throughout its length to verify that no inductor has been incorporated in the circuit. (eg. operating coils and transformer windings).

When the protective conductor is not steel conduit or a steel enclosure the requirements for the test current do not apply and an ohmeter can be used for the tests.

Testing of protective conductors comprising steel enclosures

When the protective conductor is steel conduit or steel enclosure such as trunking the following method, which requires a main supply source at supply frequency, can be used. A transformer would be needed to provide the 50 volt high current source. Commercial instruments are available which are suitable for carrying out this test.

Testing continuity of protective conductors
not part of an enclosure

When measurement of resistance of the protective conductor is required, (where no part of it consists of steel conduit or other steel enclosure) this may be carried out with a d.c. ohmeter. The following methods should be adopted;

Step 1

Connect one terminal of the continuity tester to a long test lead and connect this to the consumer's earth terminal, as illustrated on next page.

Step 2

Connect the other terminal of the continuity tester to a short lead and use this to make contact with the protective conductor at various points on the installation, testing such items as switch boxes and socket outlets.

The resistance reading obtained by the above method actually includes the resistance of the test leads. Therefore the resistance values of the test leads should be measured and this value deducted from any resistance reading obtained for the installation under test.

CONTINUITY OF CPC

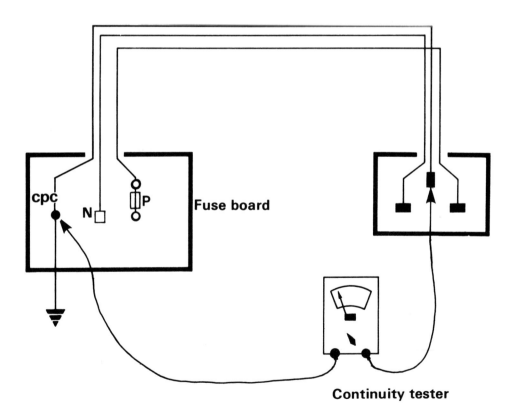

Continuity tester

Alternative Method

If the distance between the fuseboard and circuit under test involves the use of very long test leads, an alternative using the phase conductor as a test lead may used.

- Strap the phase conductor to the protective conductor at a distant socket outlet so as to include all of the circuit, and test between phase and earth terminals at the fuseboard as illustrated.

- The resistance measured by the above method includes the resistance of the phase conductor from the main switch to the point under test.

The approximate resistance of this conductor can be obtained by joining together the phase and neutral conductors at the socket outlet (at the point under test) and measuring the resistance as shown below. The value of conductor resistance is half the value obtained by this test.

The value of earth continuity conductor resistance is calculated as the initial reading, minus phase conductor resistance.

INSULATION RESISTANCE

These tests are to verify that the insulation of conductors and electrical accessories and equipment is satisfactory and that electrical conductors or protective conductor are not short circuited, or showing a low resistance (which would indicate a deterioration in the insulation of the conductors).

Type of Test Instrument

An insulation resistance tester should be used which is capable of providing a d.c. voltage of not less than twice the nominal voltage of the circuit to be tested (r.m.s. value for an a.c. supply). The test voltage need not exceed

- 500V d.c. for installations connected to 500V

- 1000V d.c. for installations connected to supplies in excess of 500V and up to 1000V.

Pre-test checks

- Ensure that neons and capacitors are disconnected from circuits to avoid inaccurate test value being obtained.

- Disconnect control equipment or apparatus constructed with semi-conductor devices. These devices will be liable to damage if exposed to the high test voltages used in insulation resistance tests.

Insulation resistance tests to earth

All fuses should be in, switches and circuit breakers closed, where practicable any lamps removed, appliances and fixed equipment disconnected. The phase and neutral conductors are connected together at the distribution board and a test is made as illustrated using an insulation resistance tester with test leads connected between joint phase and neutral conductors and earth.

The reading obtained should not be less than 1 megohm.

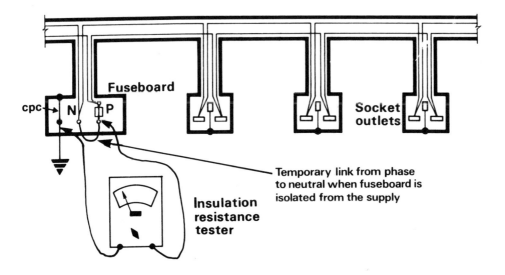

The above test cannot be carried out on TN–C (earth concentric) systems because the earth and neutral are common.

Insulation resistance tests between poles

All fuses should be in, switches and circuit breakers closed, where practicable any lamps removed, appliances and fixed equipment disconnected. For single phase circuits the test leads are connected between the phase and neutral conductors in the distribution board.

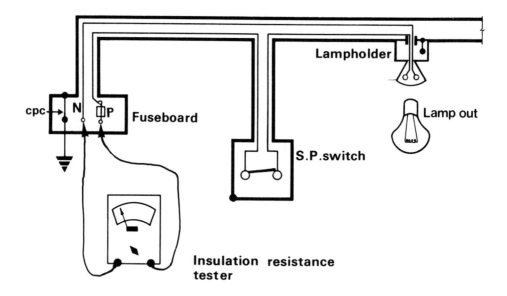

Insulation resistance
tester

Note: *Where any circuits contain two-way switching, the two-way switches will require to be operated and another insulation resistance test carried out, including the strapping wire which was not previously included in the test.*

Equipment

When fixed equipment such as cookers have been disconnected to allow insulation resistance tests to be carried out, the equipment itself must be insulation resistance tested between live points and exposed conductive parts.

The test results should comply with the appropriate British Standards. If none, the insulation resistance should not be less than 0.5 megohms).

Large installations

A large installation with many circuits has the insulation resistance of each circuit in parallel. As an example, in a large installation with many outlets, (say 50 circuits), the supply source might have an insulation resistance value to earth of 0.2 megohm. If the 50 circuits where tested individually the insulation resistance of each circuit could be 10 megohm.

Since an insulation resistance test on an installation should not be less than 1 megohm, for some large installations tested at the supply source the test would prove the installation unsuitable for connection to the supply.

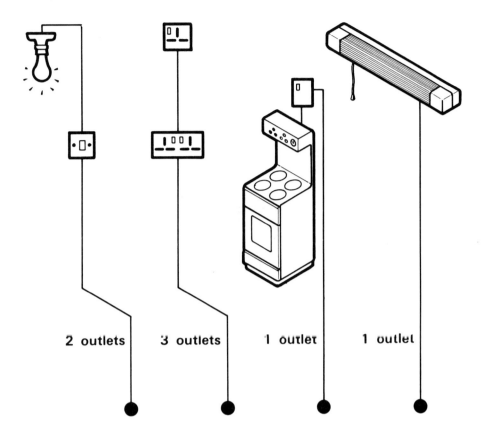

2 outlets   3 outlets   1 outlet   1 outlet

Large installations may be divided into two or more sections of not less than 50 outlets for insulation testing purposes.

The term outlet includes every switch, socket outlet and luminaire. Appliances which incorporate a switch in their construction are regarded as one outlet.

POLARITY

This test must be carried out to verify that:

(a) All fuses, circuit-breakers and single pole control devices such as switches are connected in the phase conductor only.

(b) The centre contact of an Edison-type screw lampholder is connected to the phase conductor and the outer metal threaded parts are connected to the neutral or earthed conductor.

(c) Any socket outlets have been correctly installed ie. phase pin of 13A socket outlet on right when viewed from the front.

The installation must be tested with all switches in the 'on' position and all lamps and power consuming equipment removed.

A test of polarity can be carried out using a continuity tester as illustrated.

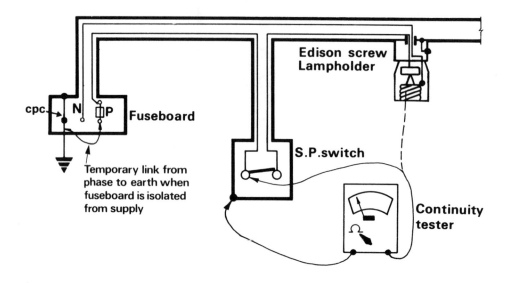

CERTIFICATION

Following the inspection and testing of an installation, a completion or inspection certificate in the form illustrated in Appendix 16 of the IEE Regulations should be given by the contractor or person responsible for the work to the client.

The certificate should be signed by a competent person.

Alterations to Installation

When changes occur in the use or ownership of buildings, alterations or additions to the electrical installations often take place. It is important that the person carrying out the electrical work ensures that the work complies with the lee Regulations and that the existing electrical installation will function correctly and safely.

A completion certificate must be made out and issued for all the work involved in the alteration, and any defect found in related parts of the existing installation reported to the person ordering the work by the electrical contractor (or a competent person as specified in Regulation 614).

Periodic Inspection and Testing

The periodic inspection and testing of installations is recommended on both the Completion and Inspection Certificates as follows :

	Not Exceeding
Electrical installations (in general)	5 years
Temporary installations on construction site	3 months
Caravan site installations	1 year*
Agricultural installations	3 years

May be extended to 3 years

The method of inspecting and testing should be in accordance with the requirements of the IEE Wiring Regulations (Chapter 6). An Inspection Certificate must be completed and given to the client.

The intervals between inspections for specific types of installations are given below:

Upon completion of an eletrical installation the electrical contractor should fix in a prominent position on or near the main distribution board of the installation, a label with details of the date of last inspection and the recommended date of the next inspection.

The notice must be inscribed with characters (not smaller than 11 point), as illustrated below.

IMPORTANT

This installation should be periodically inspected and tested, and a report on its condition obtained, as prescribed in the Regulations for Electrical Installations issued by the Institution of Electrical Engineers.

Date of last inspection ....

Recommended date of next inspection ....
